elementals

iii. water

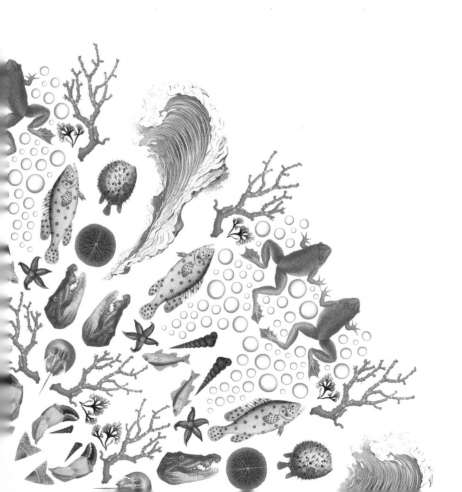

volume iii
water

Ingrid Leman Stefanovic, editor
Nickole Brown & Craig Santos Perez, poetry editors

Gavin Van Horn & Bruce Jennings, series editors

Humans and Nature Press, Libertyville 60030
© 2024 by Center for Humans and Nature

For more information, contact Humans & Nature Press,
17660 West Casey Road, Libertyville, Illinois 60048.
Printed in the United States of America.

Cover and slipcase design: Mere Montgomery of LimeRed, https://limered.io

ISBN-13: 979-8-9862896-3-2 (paper)
ISBN-13: 979-8-9862896-4-9 (paper)
ISBN-13: 979-8-9862896-5-6 (paper)
ISBN-13: 979-8-9862896-6-3 (paper)
ISBN-13: 979-8-9862896-7-0 (paper)
ISBN-13: 979-8-9862896-2-5 (set/paper)

Names: Stefanovic, Ingrid Leman, editor | Brown, Nickole, poetry editor | Perez, Craig Santos, poetry editor | Van Horn, Gavin, series editor | Jennings, Bruce, series editor

Title: Elementals: Water, vol. 3 / edited by Ingrid Leman Stefanovic

Description: First edition. | Libertyville, IL: Humans and Nature Press, 2024 | Identifiers: LCCN 2024902608 | ISBN 9798986289656 (paper)

Copyright and permission acknowledgments appear on pages 180–181.

Humans and Nature Press
17660 West Casey Road, Libertyville, Illinois 60048

www.humansandnature.org

Printed by Graphic Arts Studio, Inc. on Rolland Opaque paper. This paper contains 30% post-consumer fiber, is manufactured using renewable energy biogas, and is elemental chlorine free. It is Forest Stewardship Council® and Rainforest Alliance certified.

contents

Gavin Van Horn and Bruce Jennings
Gathering: Introducing the Elementals *Series* 1

Ingrid Leman Stefanovic
Introduction: The Essence of Water 5

Nickole Brown
Rise 11

Kathleen Dean Moore
When Water Becomes a Weapon: Fracking, Climate Change, and the Violation of Human Rights 35

Clifford Gordon Atleo
Water in My Veins: Reflections of a Coastal Indigenous Scholar 44

Joy Harjo
Spirit Walking in the Tundra 54

Martin Lee Mueller
In the Land of Five Seasons 56

Mark Riegner
The Many Faces of Water 70

Robert Wrigley
Been Ice 82

Bruce Jennings
Water Bearing Witness 83

CD Wright
Lake Return 96

Hannah Close
97   *The Big O*

Forrest Gander
106   *Sea: Night Surfing in Bolinas*

Marzieh Miri
108   *Walking with the Invisible*

Elizabeth Bradfield
120   *Learning to Swim*

Margo Farnsworth
122   *The Crawdad Quadrille*

Geffrey Davis
131   *The Fidelity of Water*

Lyanda Fern Lynn Haupt
133   *Thirst: A River's Marginalia*

Anna Selby
141   *The Rule of Thalweg*

Ingrid Leman Stefanovic
143   *Are Rivers Persons?*

Lisa María Madera
The Empathy of Rivers:
154   *Dreaming with the Río San Pedro*

Özge Yaka
Water, Gender, Justice: Women's
166   *Anti-Hydropower Activism in Turkey*

J. Drew Lanham
177   *Isle of One*

180   *Permissions*

182   *Acknowledgments*

184   *Contributors*

Gathering: Introducing the Elementals Series

Gavin Van Horn and Bruce Jennings

T hunderous, cymbal-clashing waves. Dervish winds whip-
ping across mountain saddles. Conflagrations of flame lick-
ing at a smoke-filled sky. The majesties of desert sands and
wheat fields extending beyond the horizon. What riotous conflu-
ence of sound, sight, smell, taste, and touch breaches your imag-
ination when you call to mind the elementals? Yet the elementals
may enter your thoughts as subtler, quieter presences. The gentle
burbling of clear creek water. The rich loamy soil underfoot on a
trail not often followed. A pine-scented breeze wafting through a
forest. The inviting warmth of a fire in the hearth.

This last image of the hearth fire is apropos for the five vol-
umes that constitute *Elementals*. The fire, with its gift of collective
warmth, is a place to gather and cook together, and not least of all
a place that invites storytelling. And in stories the elementals can
be imagined as a better way of living still to be attained.

The essays and poems in these volumes offer a wide variety of
elemental experiences and encounters, taking kaleidoscopic turns
into the many facets of earth, air, water, and fire. But this series
ventures beyond good storytelling. Each of the contributions in the
pages you now hold in your hands also seeks to respond to a ques-
tion: What can the vital forces of earth, air, water, and fire teach
us about being human in a more-than-human world? Perhaps
this sort of question is also part of experiencing a good fire, the
kind in which we can stare into the sparks and contemplate our

lives, releasing our imaginations to possibilities, yet to be fulfilled but still within reach. The elementals live. Thinking and acting through them—in accommodation with them—is not outmoded in our time. On the contrary, the rebirth of elemental living is one of our most vital needs.

For millennia, conceptual schemes have been devised to identify and understand aspects of reality that are most essential. Of enduring fascination are the four material elements: earth, air, water, and fire. For much longer than humans have existed—indeed, for billions of years—the planet has been shaped by these powerful forces of change and regeneration. Intimately part of the geophysical fluid dynamics of the Earth, all living systems and living beings owe their existence and well-being to these elemental movements of matter and flows of energy. In an era of anthropogenic influence and climate destabilization, however, we are currently bearing witness to the dramatic and destructive potential of these forces as it manifests in soil loss, rising sea levels, devastating floods, and unprecedented fires. The planet absorbs disruptions brought about by the activity of living systems, but only within certain limits and tolerances. Human beings collectively have reached and are beginning to exceed those limits. We might consider these events, increasing in frequency and intensity, as a form of pushback from the elementals, an indication that the scale and scope of human extractive behaviors far exceeds the thresholds within which we can expect to flourish.

The devastating unleashing of elemental forces serves as an invocation to attend more deeply to our shared kinship with other creatures and to what is life-giving and life-nurturing over long-term time horizons. In short, caring about the elementals may also mean caring for them, taking a more care-full approach to them in our everyday lives. And it may mean attending more closely to the indirect effects of technological power employed at the behest of rapacious desires. Unlike more abstract notions of nature or numerical data about species loss, air measurements in parts per million, and other indicators of fraying planetary relations, the elementals can ground our moral relations in something tangible and close at hand—near as our next breath, our next meal, our next drink, our next dark night dawning to day.

For each element, the contributors to this series—drawing from their diverse geographical, cultural, and stylistic perspectives—explore and illuminate practices and cosmovisions that foster reciprocity between people and place, human and nonhuman kin, and the living energies that make all life possible. The essays and poems in this series frequently approach the elements from unexpected angles—for example, asking us to consider the elemental qualities of bog songs, the personhood of rivers, yogic breath, plastic fibers, coal seams, darkness and bird migration, bioluminescence, green burial and mud, the commodification of oxygen, death and thermodynamics, and the healing sociality of a garden, to name only a few of the creatively surprising ways elementals can manifest.

Such diverse topics are united by compelling stories and ethical reflections about how people are working with, adapting to, and cocreating relational depth and ecological diversity by respectfully attending to the forces of earth, air, water, and fire. As was the case in the first anthology published by the Center for Humans and Nature, *Kinship: Belonging in a World of Relations*, the fifth and final volume of *Elementals* looks to how we can live in right relation, how we can *practice* an elemental life. There you'll find the elements converging in provocative ways, and sometimes challenging traditional ideas about what the elements are or can be in our lives. In each of the volumes of *Elementals*, however, our contributors are not simply describing the elementals; they are also always engaging the question, How are we to live?

In a sense, as a collective chorus of voices, *Elementals* is a gathering; we've been called around the fire to tell stories about what it means to be human in a more-than-human world. As we stare into this firelight, recalling and hearing the echoing voices of our living planet, we stretch our natural and moral imaginations. Having done so, we have an opportunity to think and experiment afresh with how to live with the elementals as good relatives. The elementals set the thresholds; they give feedback. Wisdom—if defined as thoughtful, careful practice—entails conforming to what the elements are "saying" and then learning (over a lifetime) how to better listen and respond. Pull up a chair, or sit on the ground near the crackling glow; we'll gaze into the fire together and listen—to the stories that shed light and comfort, to the stories that discombobulate and help us see old things in a new way, to the stories that bring us back to what matters for carrying on together.

Introduction: The Essence of Water

Ingrid Leman Stefanovic

Remember that, my child. Remember you are half water.
—Margaret Atwood, *The Penelopiad*

"Oh wow! Can you hold my hand? Please."

I was seven years old, standing for the first time overlooking the enormity, the immense fecundity of water and power called Niagara Falls. Reaching toward the security of my mother's hand, I was hoping to stabilize myself in the face of such overwhelming fluidity. In dreams many years afterward, I would feel myself succumbing to the vastness and depth of the flowing waters, roaring, cascading, plummeting over the escarpment. It is my first memory of water as elemental.

To view water as elemental—what might that mean? Certainly, as Margaret Atwood notes in *The Penelopiad*, we are constituted in large measure by water. Our very being depends on it. That fundamental need for water defines our very existence.

But there is more. Lived experience brings with it a multidimensional, kaleidoscopic array of visions, narratives, and worldviews around the element of water. This book presents some of those diverse stories in the hopes that the reader will begin to ponder their own traditions, perspectives, and aspirations around this vital, existential element of life.

In many ways, the place of water is very much taken for granted. As I reach for my morning coffee, I forget that, on average, 140 liters of water have been used to grow the beans that constitute my one cup of morning brew. My water-efficient showerhead draws 9 liters of water per minute. And the single pair of jeans that I throw on has used at least 7,500 liters in the production process—as much water as the average person drinks over seven years![1] Water quietly, invisibly, makes its way into our lives, so often without our explicitly attending to it.

At the same time, many struggle in the face of its absence. A staggering 771 million people—one in ten—live without access to safe drinking water. And 1.7 billion people—one in four—lack adequate sanitation.[2] Every day, four thousand children die as a result of waterborne diseases.[3]

Such numbers are hard to contemplate in anything other than an abstract way.

And yet that is exactly what we must do if we are to care for and nurture the planet upon which we depend.

This book presents an invitation to confront water in its lived essence, as do the companion volumes on other elementals. More than simply a commodity or a series of statistics or a technical requirement for human development, water holds greater promise when we see it as elemental to our existence. The authors in this volume aim to shed light on just such a promise.

We begin with Nickole Brown's moving essay, "Rise." By describing her experience of living through a sudden flood, Brown captures so much of the commanding presence and magnitude of water that I felt as a child in Niagara. As she notes, it is one thing to read about such moments or to view pictures in the news. It is an entirely different matter to be confronted with rising waters that threaten your lived sense of security, stability, and safety. Brown reminds us that, as much as we seek to control water, ultimately, we are beholden to its ways.

The immense presence and power of water are presented from a different perspective in the following essay by Kathleen Dean Moore, who takes us on a journey through the violence of fracking. Her essay is a moving testimonial to water's authority as it disrupts a balanced, sustainable search for energy. Moore calls us to confront not only water's immensity and power but also our responsibility to seek a more sensible, caring way of living.

This message is then picked up by Cliff Atleo, who invites us to "care for the water as it continues to take care of all of us." He illustrates how his life, so intimately connected to his homewaters, influenced his research and current role as an academic in western Canada. He confronts the challenges of Indigenous groups who acknowledge the importance of mineral extraction as part of the new economy while maintaining a reverence for the waters affected by modern industrial development. Atleo raises the specter of environmental racism, which insidiously affects public policy and threatens Indigenous waterways. And he reminds us, in a poignant story, about how the simple act of eating canned salmon while in a foreign place helped to bring him home, back to the land of the "saltwater people."

The Indigenous voice arises once again in the poem by the twenty-third US poet laureate and member of the Muscogee Nation, Joy Harjo, who opens the door to Martin Lee Mueller's own walk through the waterworld of rainy bogs. In this chapter, he describes how, within only a few days, the water level of Sooma—the land of bogs—can rise by three or four meters, "turning the network of raised rainy bogs into an archipelago of spongy islands in a seasonal ocean." Contemplating such a presence of water, Mueller reflects on the way thinking itself ultimately can be seen as "fluidified." He invites us to reflect on water as never a static *thing* but rather to try to think that "every 'thing' may be a thinging," the river "rivering," the child "childing." Our English language tends to reify the world, but when we take our cue from water, we acknowledge the elemental dynamism and temporality of life itself.

This theme of fluidity is once again taken up by Mark Riegner, who explores the many faces of water and the hydrological "flowforms" of Theodor Schwenk. Water "appears to us in many guises" as liquid, gas, and solid ice—a reflection captured so poignantly by Robert Wrigley in his poem "Been Ice."

In each of these literary moments, we find ourselves in the presence of what Bruce Jennings describes in his essay as water "bearing witness." Jennings shares his personal stories of the healing power of water, specifically when, as a child, he contracted poliomyelitis and recovered through the careful and caring process of hydrotherapy. Fishing for bass becomes a story of family, and then we are graced with intimate recollections of water memories and his wife, Maggie Jennings, who passed away in 2020. The chapter presents musings of water at the intersection of remembrance, meaning, and love.

The link between water, love, and sensuality is then further revealed in CD Wright's poem, and in Hannah Close's essay that follows. As the origin of earth's earliest life forms, "the ocean is eros writ large," writes Close. That perspective is carried forward by Forrest Gander's poem "Sea: Night Surfing in Bolinas."

That this volume aims to immerse us in the rich, multistoried presence of water does not always mean that such presence is empirically visible, as Marzieh Miri reminds us. She describes how she sought to explore a hidden river, buried over the years beneath the urban environment of Toronto, Canada. She encourages us to learn how to look and listen "to what cannot be immediately seen or heard—a lost piece of nature in the city, retrieved and represented through an installation of video, cyanotypes, and sound."

That water invites an adventure of learning is further explored by both Elizabeth Bradfield in her poem and by Margo Farnsworth, who encourages us to learn from the practice of biomimicry. Farnsworth illustrates through stories and personal reflections how water as an element is both critical to our survival but also "a substance cradling the life-forms from whom we could and can learn." Her essay serves as a reminder to remain true to what Geffrey Davis describes in his poem, as "the fidelity of water." The theme of thirst, touched on by Davis, is further contemplated by Lyanda Fern Lynn Haupt, who reveals how "authentic love of water can be formed only through the crucible of thirst"—ultimately an elemental thirst "to continue, to simply go on." Haupt beautifully concludes how her "own deepest thirst twines with the dream of these birds, these fish, this river, this forest, this earth."

The theme of rivers is then contemplated by Anna Selby in her poem, "The Rule of Thalweg," and by the volume editor, Ingrid Leman Stefanovic, who asks whether rivers can meaningfully be referred to as "persons." Investigating the global phenomenon of assigning rivers legal rights, Stefanovic wonders whether there is more than legal utility to this movement. In the following chapter, Lisa María Madera expands on Stefanovic's question, showing us how we can meaningfully take a stand to preserve our rivers and relate to them as living subjects, focusing on the example of Ecuador, one of the few countries proposing new rights for the natural world.

The final essay, by Özge Yaka, provides another example of environmental activism around water. Drawing from the example of activism in Turkey, she reminds readers of the importance of taking a stand to protect our waters as more than just an academic exercise of assigning moral value to our waterways. It is vital that we each do our part to preserve, protect, and nurture this elemental force of water, so essential to our and our planet's future.

J. Drew Lanham's poem "Isle of One" closes out this collection. Reminding us that "there's no away that's far enough away any more to ever be insulated or safe," Lanham leaves us with a sense of responsibility in caring for our watery worlds, together with a sense of hope.

The Renaissance painter, engineer, and theorist Leonardo da Vinci proposed that water is "the driving force of nature."[4] Through these wonderful water stories, the reader comes to know and to feel, in a visceral, moving way, how this is the case. As you make your way through your day, I invite you to attend to the presence of water. It is not always as awe-inspiring as it was for me as a child at Niagara, but it is always there, nourishing us, protecting us, inviting us to live in a manner that preserves it and the world in which it makes its way. The journey of life is itself a journey of water. May these stories open the possibility of your living with water in a more caring way.

notes

1. See the "UN Launches Drive to Highlight Environmental Cost of Staying Fashionable," *United Nations News*, March 25, 2019, https://news.un.org/en/story/2019/03/1035161.
2. On the water crisis, see the website Water.org: https://water.org/our-impact/water-crisis/.
3. Jessica Berman, "WHO: Waterborne Disease Is World's Leading Killer," *VOA*, October 29, 2009, https://www.voanews.com/a/a-13-2005-03-17-voa34-67381152/274768.html#:~:text=According%20to%20an%20assessment%20commissioned%20by%20the%20United,of%20diseases%20caused%20by%20ingestion%20of%20filthy%20water.
4. Cited in Laurent Pfister, Hubert H. G. Savinije, and Fabricio Fenicia, *Leonardo da Vinci's Water Theory: On the Origin and Fate of Water* (Oxfordshire, UK: IAHS Publishers, 2009), vii, https://water.usask.ca/hillslope/documents/pdfs/2008/08-05%20SP009.pdf.

Rise

Nickole Brown

*Oh, my proud children, wherever are we going on this mighty
river of earth, a-borning, begetting, and a-dying—the living
and the dead riding the waters?*
—James Still, *River of Earth: A Novel*[1]

You watch and you watch and you watch.

So you watch and think you know. Yes, you scroll and scroll,
just like I once zipped through the channels as a kid—before I had
a phone, before there *were* phones like now—and I was watching
then as I still do, the false fire flickering across my once-young face
a static blue, especially blue these days as I hardly stop watching,
can't seem to turn it off.

So someone says *flood*, and you think, *Yes, I've seen floods*. Because
of course you've seen floods. I know I have—on the news and news
feeds, on bigger and bigger screens, some in bars and others in
airport gates, one even in a nail salon with a technician kneeling at
my feet—flood after flood after flood after—so many now they slop
and writhe as one.

But really what you think is a flood isn't *a flood* but *pictures of a
flood*, isn't the real deal but its representation, not the signified but

its signifier, a complicated experience made gestural, ready to be consumed. Nonetheless, you keep watching, keep clicking through:

Not cars floating but pictures of cars floating. Not waves choking the slender neck of a stop sign but footage of that turbid water, of that red octagon heeded no more. Then, almost always: a woman clutching a drenched and quivering dog. Then, almost always: a man behind her, life-rafting what little he can alongside him in a plastic storage bin.

Underneath it all, predictable captions: *salvage, rescue, damage, toll, rise.* You read the words, understand them, or at least think you do. But then you're distracted by the buzz and ping of all that's incoming. Little red circles and little red bells, one notification and the next, urging you to move on.

Yet that last word you read—that word—*rise*—it rises and rises up the floor within you, brings on a subtle quease, like you're riding an elevator descending too fast.

I say this because now, I know.

Because now I've lived through. Now, I understand: specifically, a flash flood, in the mountains of Eastern Kentucky, July 27, 2022. Among the forty-three who died: ten in Breathitt County, two from Clay County, and, in the county in which I was in—Knott—nineteen lost their lives. Among them were four children from one family who clung to a tree until they couldn't no more. Later, I heard that all four were buried in one casket, returned to the earth the same way they died—all together at once.

I was there, in Knott County, teaching for the Appalachian Writers' Conference at the beloved Hindman Settlement School, having myself a good time eating tomato pie and talking poetry. I was listening to old-time fiddles and sipping bourbon on the porch after supper, and when I strolled back to the dorm, the night was thick with katydids, their sawtooth wings chirring a song that to me *is* July itself.

The flooding began just about then—midweek—a Wednesday—well into the night.

But no, that's not true. It began well before: all that week since we arrived, it had been raining. But not too hard, not really, or so it seemed. Again, I was having myself a good time. Like nearly everyone else, I hardly paid mind to the rain at all, even if it was raining hard.

Forgive me then: I want to show you what a flood is, what it can be. Because although it may not be raining where you are, at least not now, all of us, in one way or another, need to know, need to prepare.

First, understand the illusion of home, how we believe we're safe because safety is something we need to believe.

Consider the room where I slept: hardly my home, but still, a kind of home—the AC hummed, a clean quilt tucked to my chin, my little dog who'd traveled with me curled and softly snoring at my feet. Across the room, one roommate was sleeping sound, her hands moisturized, her eye mask in place. On her nightstand, the night's glass of water, her bedside reading closed with a bookmark marking where she might pick up tomorrow, confident tomorrow would come.

So when my phone crowed its first weather warning, I wasn't jolted awake, no. I was annoyed. I half-read the word *flash flood* and bemoaned how technology never let me be, thought how silly that something happening far from here should try to rattle a moment peaceful as this. I silenced the alarm, fell back asleep.

About an hour later, our other roommate came in for the night, and though hesitant to wake us, she stood dripping in the doorway and said, *You guys won't believe this, but the creek's rose. My car—it was down in the lower parking lot—and the water's up to my steering wheel.*

Would you believe me if even then I fell back asleep or at least tried to? That all three of us did?

Yes, we might have talked about her poor car or made a joke about the Troublesome Creek living up to its name, how Appalachians hardly mince words if they can help it and no doubt christened that deceptively benign waterway with a suitable adjective. But we weren't worried, no. We agreed: her car was a good distance from us down a steep hill, and besides, it wasn't coming down all that hard. Because, yes, that's the tricky bit: the deeper you're tucked in, the more the illusion persists.

What's funny is even when I *did* get up—after I'd tossed a long while before reluctantly deciding to move my own car to higher ground—I felt silly, feared one of the students might see me traipsing across campus with my pajamas ballooning from my boots, might see me with no bra.

It was only when I got back to the room did I realize the water had come ever closer, did I begin to sizzle with fright. It was then did I begin to wake. From a distance, the black shape of my roommate's white car—not just flooded but now floating, the hazards flashing as if turned on by a ghost ready to drive across the river Styx.

But even then, I felt ridiculous. I was embarrassed, sure I was being a drama queen. I made fun, said I'd been reading too much Anthropocene poetry, said if there was a national emergency, it was just me flash-flooded with hormones, said if they'd humor me and gather their things—*just in case, just in case, I know this is silly, but just in case*—I'd skip out of this dry county tomorrow to a wet one and get them a bottle of their choice. We had a laugh at that— calling the county *dry*. For a minute, our talk mother-eased me, almost had me quit my fretting and go back to bed.

One roommate teased, asked me if I had a catastrophic childhood or something, and I teased back, said, *Well, I told you I grew up in Kentucky, didn't I? Y'all will just have to forgive me for expecting the worst.*

We kept laughing, but all the while, a deeper me began to quake, recalling just how Betty Crocker of a comfort was my childhood home—how sweet and warm the cornbread from the oven, how white and warm the sheets from the dryer—and just how all of that made what happened down in the basement when Mama wasn't looking seem so unreal, so impossible to believe, even now. It stung, that open-handed slap of memories. It made me gather my things even faster.

Before we were done zipping our bags, water snaked under the front door, fast, in long and slender tongues. It was the color of sickness, the weak yellow of bile when a body has vomited every-thing up but can't seem to quit.

Unable then to open that door and step outside, we lugged our be-longings in through the building, through the dank hallway and up the stairs, the water writhing behind, chasing us. The electricity stuttered, then stuttered again and quit, so we sloshed through the dark toward the red glow of Exit signs.

Once upstairs, alarms beeped on and on, the way they do when someone in a hospital dies, the beeping incessant and unkind until a nurse comes to give the final word and turn off the machines. I flipped every switch I could to make it quit, and when it finally did, I expected silence.

Instead: a groaning, deep and Titanic, like a great ship going down.

Because below us, water fast and faster still. Below us, water gurgling up pipes, water bursting glass doors, having its way with the sizable archive stored below, a hundred-plus years of Appalachian literature and photographs destroyed. Below us, our beds lifted and bobbed toward the ceiling. A yellow dress I had hung on the back of the door for the morning and forgot wicked water up to the neck.

The dark made the flood seem loud, and in the dark, it hissed and bubbled. Creaked and blasted. Moaned and sighed. A sound like the dark boiling, like the boiling dark. A sound sounding out a name I will never forget. I tied my dog's leash around my waist, reached to pet her trembling back, felt grateful the door that led outside this floor was still on dry ground.

And that was our first escape from what the Troublesome brought.

So, then, know this: how water works: how it comes from above as from below.

You see, before last summer, I had what most do of floods: an aerial view, which is why perhaps I thought the velocity was that of rain, falling from above. Yes, aerial—just like on the news—the view from the camera aperture of a drone, seeing as a reporter does, wearing noise-canceling headphones, the helicopter's battering *chop-chop-chop* tossing his hair into a news-at-eleven panic.

But no one truly sees a flood from up there, not really. No, nothing, except for an exhausted cloud or a kiting-high plastic bag. No, no one, except for maybe a bird frantic over her gutter-drowned nest. Except for maybe God, if you believe in *up there*.

Instead: know this: water: how it moves: not only from high to low, but from low to high. How if there's enough of it, it wakes, rises from where it rests, rises from the table from which we take our drink.

And so, as from above as from below: first the floor drains spit up what they were made to choke down, then the toilets and sinks. And the flooring: first just slightly cool under our feet, a dampness easily dismissed. Then, a wet stain phantoms up. Finally, all of it—the rugs and carpets, the tiles and floorboards—they, too, lift and float.

I say this because the rain that week was comforting, seemingly benign. At times, it was a snuggle-down, book-reading mist. Or an easy, window-glazing kind of drumming. I mean, consider this: my roommate—even after her car floated away—she curled into the sill, said, *Rain like this always soothed me. Ever since I was a child, it just makes me feel calm.* And the other roommate, she later wrote to say the rain that evening didn't look menacing to her, just odd. *So I took a video of it,* she said, adding, *But who takes videos of rain? Well, I'm sure lots of people do, but it never occurred to me before. A form of praise, I think.*

But when it floods, the word *water* tendrils, morphs into *waters*, that single letter—that *s*—makes water plural, that singular noun makes not one unified water but water from all directions, across from rivers and up from wells and down from clouds—*water* made *waters*—running down mountains and melting off ice caps, making creeks of ditches, rivers of creeks, lifting the good earth beneath our feet and making it slide. *Waters,* rising and rising, without cease, like a bathtub running and forgotten, like a bath running for someone slumped within it, their wrists slit.

Waters sounding somehow more poetic, more holy. *Waters* like a body letting go of its water, muscle made mud, eyes made mud, skin and hair and organs—all mud, mud, mud—a body returning to the thing from which it was made.

Waters, sounding downright biblical, ready for a baptism maybe and given power, which is what water is.

Long after, I'll take a sip of coffee, think what if this is not coffee but water that remembers it has passed through something, that it's been changed, stained by me and a scoop of grounds? What if this were not *coffee* but *water*, which truly, it is? What if it were not *water* in my mug but *waters*, what does that mean? I take another sip, think this is the kind of shit that keeps me from getting anything done, the quivering I can't seem to quell, even months after that flood. I shake it off, answer another email. But still, a part of me never stopped wading through that ceaseless night.

Because a flood, it will lodge in your body; it will change those who wade through, will call to task what your eyes think they know, will have your ears and nose summon the jagged rest.

Thus, the smell: high electric, an ion-charged whine, a seething scream of petrol and piss, a way-up-in-the-head stench, migraine-tinged, lit by the kind of fluorescents in places where you'll see things you don't want to see. Worse, the smell is not just sewage and gasoline but is made discordant by a smell you love: petrichor, that earthy goodness of summer showers, the microbes long kept in the soil rising up again.

What I mean is the smell is part earth, part what we have done to the earth.

What I mean is the smell is part miracle that made and sustains us, part what we have done to destroy that sacred covenant, what we have likely done to cause this mess. Once you breathe it, it will reside within, your lungs watermarked like the walls of a once-flooded room now dry but under that fresh sheetrock a discolored line that never forgets.

Worse, it's a smell made visible by what's caught on the fence line: plastic bottles of detergent and plastic bottles of antifreeze and plastic bottles of hand sanitizer and plastic bottles of drinking water looking ironic as hell floating in all that water. And there, in the limb of a tree, even a plastic bottle for a baby, still somehow half full of milk.

Later, it's a smell doubled by things you can't first identify or even fathom that come to knock down the fence and the power line and whatever else is in the way—a floating parade of feed troughs and dumpsters and pickup trucks, porch furniture and couches. Later, tool sheds and trailers, the wet rag of a dead raccoon, the limp hose of a dead snake. Then, a thing that makes you laugh a little but in shame: an above-ground swimming pool, its beach ball pinwheeling green and pink and yellow in the fetid rush.

Further down the road, there's a house lifting from its foundation that crashes into another. They say those two homes then lift together and rush into another—three homes made gruesome dominoes that freight-train into an eighty-something woman who lived her natural born in these mountains until that night.

And know the flood won't just lodge in your body. No, your mind, too: it will split.

Because not far down that first dark hallway, I could feel it happen: While a part of me was ever present, another was far into the future, remembering already what was happening now.

Yes, you will split:

A part of you will be pure amygdalae, keenly aware, all your senses tinged with adrenaline, but at the same time, another you will cling all at once to the past and the present and the future, desperate to make sense of what's happening before it's even over.

Put another way, you'll scramble to figure how, if you live, you're going to tell this story.

No, that's not quite right: you'll feel if you *don't* figure a way to tell this story, you *won't* live.

No, let me try this again: you'll tell the story to yourself in past tense as it happens, minute by minute, each moment already translated into the future, already a story you've known your whole life.

Consider the walk to the next shelter we could reach at the top of the campus's main hill—an old house made into a dorm. In my hand, a book light no bigger than a fairy's lantern to illuminate each step, my boots disappeared in that khaki-colored rush, my feet alert to every tug and vibration, feral with the pock and tap of every small rock and stick charging past.

At the same time, before I got to the door, I was already at the door, had already locked and loaded my line, saying, *Have room for a climate refugee?*

It was meant to be a wry joke, the kind of line I'd want the actor playing my part in the movie to say, a commentary on this terror weather told in a lighthearted way. But then I thought, *No, how arrogant, how Little Miss Privilege of me... A frightening evacuation from one building to the next hardly a refugee makes. Besides, you don't even live here; your home is a hundred miles east of here, safe and sound.*

I rewound the film, edited out that scene, shot the footage again. I said that line then unsaid it, all while walking through the door I had not yet reached.

So I opened the door and open the door and will open the door, all at once; I coaxed the verb *open* to open to all tenses together and do just that to the door.

When I do, the door opens to a silent and sleepy-time dark—there is no one to even receive anything I could say. They are all still in bed; none of them even know what has happened. None of them know what's happening now. They don't even know half their vehicles have washed away.

Hello??!!? I said, drawing out that knock-knock-anybody-home vowel of *o*.

Hello??!!? I say, making a fist and using it on one door then the next.

Hello??!!? I will say, again and again, into that dark forever, waking each person, telling them to rise.

Eventually, all of us gather together in the pitch black with no idea what to do as great waters form around us and roar past. We tremble to think the steep incline above us will slide, knowing there is nowhere else to go.

I feel guilty—there is nothing any of us can do. I should have let them sleep. It's only when one dreams that all tenses should operate like that at once.

Here then is another thing to know: when it happens, when a flood comes, you won't act the way you think. You might not even notice the flood has washed the you that was you away.

See then the motley huddled dozens of us, together in the dark:

One crumples and cries, murmurs something about a fire. Another stumbles backwards, sloshing liquor to the floor. Another blasts the room with a machine-gun fire of *fuckfuckfuckfuck* then chucks in a grenade of *goddamnit, Appalachia, goddamn!* Another runs outside to try to save her motorcycle and returns defeated, saying she had just paid for it in full. One walks around with a Sharpie, telling us to write our names on our arms just in case our bodies need to be identified. Someone else paces, another falls asleep snoring upright in a chair, someone else says, *Well, hell's bells, I'm going to have a smoke then, because what difference does it make now if I quit?* Another distributes blankets. Another gathers the blankets and puts them up. Another passes the blankets back out again. One stares as a hawk about to strike prey; another stares as a horse dozing in the sun. Yet another tells us of a flood she escaped as a kid, how her mama in a frenzy to evacuate gathered only three things: money from the dryer, a bunch of bananas, and her girdle. I have a good laugh at that, deeply thankful for the kind of humor Kentucky brings.

And me? I churn petty anxieties, chide myself for what I left behind in the room: my dog's treats, my dental floss, that yellow dress. I open my bag, check to see if I remembered to get my hoodie, my notebook, my computer, my wallet, my phone. I let my dog off leash so she can perhaps comfort people, then I panic, worried I won't be able to find her if the hill above gives way. I clip her in again, double-knot her leash to my waist. I check for my hoodie once more. I check for my hoodie again, my notebook again, my wallet, my phone. I check for my jewelry and then put my jewelry on, careful to secure the heavy bracelet my mama bought me when I was but twenty-one, finally facing those dank early memories I had tried so hard to erase. Instead of taking my own life that day, I went shopping with her, and for over two decades, I've worn that bracelet, wielding it like some Wonder Woman kind of cuff. I put it on my right wrist, figure it will either be heavy enough to keep me from floating away or that's what will identify my body when they find it, because I'm surely not going to take a Sharpie and magic marker my name to my forearm.

More and more people from the lower buildings arrive, filling the crowded dorm so much that we spill out onto the covered porch. A cat yowls from her crate. A little wet dog yips and shivers in the bathroom, and when my dog sees him, it's as if she couldn't care less about meeting other dogs anymore and pays him no mind. Again, I tighten her leash around my waist, figure if we're carried off at least we'll be together. Figure if I need a length of rope to tether the two of us to a tree, I can use that.

Not much later, a friend leans in, says, *You know, the land above us is forested; the ground above us should hold.*

It's then I stop my spiraling, take the needle off the record going round and round in my head.

Much of the land here has been logged completely, he adds. *But the trees above us should hold. But we won't know until morning.*

I can't remember saying anything in response, but to myself, I thank the trees for holding and thank whoever decided to leave those trees standing, especially knowing just how hard money is to come by in these mountains and for how much money such trees could be sold. Then, from deep down comes a scrap of lines from a half-remembered poem:

Earth loved more than any earth, stand firm, hold fast; / Trees burdened with leaf and bird, root deep, grow tall. Mister Still—as folks still call him around here—he wrote that.[2]

Because yes, how much land here been logged? I've never lived in Eastern Kentucky but still: all those stories about my great-grandmother feeding lumberjacks out this way, how she'd fry up twenty-one chickens at a time for those men, just for lunch, how she lived in poverty that ground the word *grinding* to dust. And just how many trees did those hungry men cut? How much coal did they crawl into the earth to pick and haul? How many trains did it take to carry all of that far away from here, to build and heat cities full of people who would look down on this place?

Yes, how much money those trees could earn, how much even one tree might cost. How many lives huddled below those trees that night.

Even now, I repeat Mister Still's words, like a mantra, again and again and again: *Earth loved more than any earth, stand firm, hold fast; Trees burdened with leaf and bird, root deep, grow tall.*

Again, say it with me: *Earth loved more than any earth, stand firm, hold fast; Trees burdened with leaf and bird, root deep, grow tall... Earth loved more than any earth, stand firm, hold fast; Trees burdened with leaf and bird, root deep, grow tall.*

More than any earth. Hold fast. The trees, burdened. The trees, left to grow, to root deep. Earth, firm, deep, more than any earth. Bird, loved. Trees, loved. Earth, loved, hold, hold, hold. Toss the words from those lines around any way you want, and you will still find a prayer there worth praying.

Know too there will be an *after* that comes after. There will be more to follow; the story won't just end after you're too tired to tell the story to yourself. No, it will go on and on like a story you don't want to watch anymore but can't turn off.

Which means even though this night will end, the story will not. Which means instead of morning like you've always known, the first light will reveal not the return of day but a day turned in on itself, a day torn from the one that came before.

Which means at first light will be four ducks in front of the dorm, and though amused by their waddling, you realize they're clearly domestics and are searching back and forth and back and forth to take shelter inside a barn that's far from here and likely no more. Eventually, they'll stop their panic and will curl exhausted next to each other, their fresh-snow necks tucked into their fresh-snow backs, an unreal spot of brilliance on that mud-slaked hill.

Which means too that at first light you'll walk to what's left of downtown, amazed the little convenience store—the aptly named Mi-Dee Mart—is still there. You'll stand at the edge of a furious brown river that yesterday was Main Street. Next to you, a man smokes the way working men you've known all your life do—arm down, hiding their cigarette at their side, using their palm to protect the glowing cherry from the wind, smoke rising from their open fist. *Never in all my forty-five years have I seen such a sight,* he says, not to you exactly but meant for you to hear. Next to him, another man, smoking much the same country way, replies, *No, never, not in all my twenty-five years.* And yet another, exhaling a long time as if he's pushing smoke out of his lungs though he's not smoking a thing, says, *No, never. Not in all my seventy-three.*

When you return to the dorm, it's decided that those who still have cars to speak of should try to get out of there while they can. Besides, more flooding is predicted to come, and the settlement school needs all the housing it has left to help those who live there. You'll never forgive yourself for this, but you do as you're told: you leave quickly, without so much as a goodbye. You flee without staying to see the waters recede, you take off without helping to haul so much as one box of precious archives out of the mud. You don't even stay long enough to see if the owner of the four little white ducks will come and take them home.

And here is where you'll have to forgive me for speaking in third person, because this was *me*, not *her*—I know—but the evacuation is too much to bear with blatant capital *i*'s as if *I* were there. Because, yes, telling this next part in first person would be a lie. So I'll tell you the rest in third because the me that is me was, by then, long gone, because driving away that day was a stranger—a stranger not *to* me but a stranger *of* me—an artificial intelligence that felt nothing as she listened to what the me that was left of me tried to say.

Because all that day, she tries and tries to get back home, but there's more mud than pavement, more water than road. She hydroplanes through one intersection then the next, and when the road is impassable, she turns the car around, hydroplanes through the same intersections all over again. She rides the rumble strip on the side of the highway, and later, she turns around yet again and rides that same rumble strip once more, this time on the wrong side of the road.

The gas pumps will be closed; the gauge will nearly tick down below *E*. There's no place to stop for food. She'll come to one dead end and then the next and the next, not sure if the water gets deep enough if it's better to shut the engine off or gun it or what. She'll pass one destroyed home after the next; she'll dodge dislodged trees and stones; she'll dodge mattresses and clothes and other cars not as lucky as hers.

Right outside the town of Hazard, she'll pass a brown dog twisted into a pile of meat, and unbelievably, a swallowtail will be drinking from the side of his wet neck. She'll see that dog again and again; I even see him sometimes today—the twist of his open muzzle, the heap of red, that flitting yellow pumping its bright wings.

Unable to find a way east over the mountains to get back home, she'll finally turn around, head west instead, away from where she doesn't belong now anyhow, away from home.

Exhausted, she'll stop in central Kentucky, and there she'll find an empty room empty of everyone except her dog and herself. The room will be basic and cheap; it will be blessedly clean, blessedly dry. When she sees the white of sheets crisp on the bed, she won't weep, though she wishes she could. She won't bathe. No, she won't even take off her shoes. She'll slump her filthy body into that clean bed and won't even remember falling asleep.

The next afternoon, she'll wake. In the mirror, she and her dog both will be slaked with dried mud that comes off in flakes and drifts; there will be a fine dusting of red-brown on everything. Her pajamas will still be tucked into her boots and she still won't be wearing a bra. She'll wonder at what point she went from being ashamed the night before last to not noticing how she was dressed at all.

On the counter is the kind of coffee machine with a filter tray made of plastic so disposable it melts and falls off when the hot water runs through. Little plastic bags of napkins and little plastic bags of plastic stir sticks. A tiny bag of powder to turn her coffee pale and pretend it's cream. A tiny bag of sweetener with a cancer warning on the side. A small sign that states the grounds are 100 percent rain-forest select. She makes a cup but can't drink it; the weak brew's the color of that water she just left behind.

She'll pour out the coffee. She'll shower, crawl back into bed. It's there she'll stay for two days.

Finally, know too how once you turn on the television to watch the news, how what you've just survived will somehow seem less real, even to you, how it will slide back into the familiar trope you've seen so many times before. Know how much work it is to hold on to what the body knows, how the media will try to tell you how to feel in between commercials about pills for depression and pills for blood pressure, commercials for pizza and burgers and under-twenty-minutes-or-less-easy-breezy meals.

Worse, as the world is made as it is now, it also demands that you live two lives—one in-person and another online—so you pick up your phone and scroll through to find another foul current, this time made of words:

These people got what they voted for, one says. And yet another: *We should just let them swim.* And another: *What are those houses doing there along the river in the first place?* And another: *Maybe it's God's punishment for being a bastion of ignorance and regression.* And another: *I've never seen so many banjos floating down the river.* And another, and another, and another, so much so I wade through another long darkness in a different kind of flood.

I also watch clips of people from the drowned counties, hear in their talk the accent that once thickened my own tongue, the kind of talk that somehow others think it okay to imitate with a *Beverly Hillbillies'* yee-haw, that has them hum the notes from *Deliverance* in case I didn't get the joke. Yes, even my most educated and liberal friends make redneck jokes to my face and don't think a thing of it.

We didn't have no time to get nothing out, one woman says in one clip, her heartbreak of a double-negative shaking its head *no* and *no* and *no.*

We're not gonna let up til everyone is accounted for, says another, and I think of how I once cut my teeth on the word *kin,* far before anyone ever uttered the word in any academic circles, how I was brought up to think blood thicker than any water, even floodwater thick with blood itself.

Later, a journalist will call to get a statement from me on the phone and will say what he can't say in print: never in all his years of reporting flooding in Eastern Kentucky has he seen anything like this. Regardless, if he says the words *climate change* in his article, what kinds of hate mail he might get.

Somehow, it makes me recall an old joke I've heard more times than I can count, one about a hillbilly stuck on his rooftop in a flood, crying out to Jesus to save him. Soon, a man in a rowboat comes by, says for him to jump on in, but the man refuses, says he was praying so hard that Christ Himself would come down from the cross and save him. Then a motorboat comes by, offers the same, but the man says no, that he's righteous with a snakebite kind of faith, says he sure doesn't need the likes of any help from mankind. Finally, a helicopter buzzes down and the pilot shouts for him to grab a rope and be lifted to safety. Again, the man refuses, raises his arms, speaks in tongues instead.

Perhaps you've heard that one already? If not, maybe you've heard a simplified version of the same? It's well known enough, and more than one person told that joke while we waited for the sun to rise the night of the flood. Either way, I bet you can guess the punchline: The man drowns and goes to heaven. Once there, he asks God why his prayers weren't answered, why he just let him die like that. To this, God replies, *Well, I sent you a rowboat and a motorboat and a helicopter... What more did you expect?*

What I can't help wonder though is this: What if there's much missing from the joke? What if that first would-be rescuer, the one in the rowboat, made fun of the way the man talked, figured that helping him was a charitable thing in a better-than-thou sort of way? And what if the owner of the motorboat was trying to coax the man off his roof in exchange for the rights to his land, figuring there was good seam of coal gleaming under all that brown water, that those old-growth trees would be that much easier to cut after they died back from the flood? And what if the helicopter was just another government handout, there to lift the man up for a spot on the news before ditching him in some arena crowded with evacuees where he'd have to work that much harder to even think of finding his way home?

Worse, what if none of that was true—what if each person who offered help was as good intentioned and kind as they come—but the man, having sprung from generation upon generation of exploitation and ridicule simply couldn't trust the likes of any of them and instead put all his faith in God?

This, of course, is the kind of pissed-off thinking that gets me in trouble, that keeps me from taking pleasure in a simple joke.

So now, I know. And maybe now you know or at least know more than you did.

So come with me then, back to the moment before it began, before I didn't know, back to that *before* to which I return, again and again: to that last night I had in Hindman, in the eastern part of the state, at that beloved settlement school there. Again, you know the story by now: It's Wednesday night. It's raining. Of course it is, but we barely notice.

Yes, it's raining, but it's not a thing that worries us, at least not yet. Instead, we listen to katydids thicken and texture the air—that sound that *is* Kentucky in July itself.

If you would, then, slow with me a spell. Let the song enter us; allow that sugared darkness into you, that sweet dark that frightens and soothes at once. Because in that insect song is something that says despite all the words we read and words we spoke and words we heard today, we're still animals just as my dog is an animal and the katydids are animals though my dog is sitting at my feet and looking up to ask us why her walk has stopped.

Please: stay with me, just a beat longer, not so much *listening* to the katydids as we *drink them*, with our ears, and let me tell you an Appalachian tale I half-remember—about how theirs is a call-and-response song of *Katy did, Katy didn't... Katy did, Katy didn't...* something about a girl named Katy who killed somebody—maybe her sister, maybe a man she loved, maybe both—but whoever it was, the katydids way up in the branches saw everything like they always do and now have something to say.

We maybe laugh then about stories from these mountains and how they almost always involve some hideous crime of passion—*nothing better than a murder ballad sung by bugs,* of all things—and we'll listen to the insect jury debating on and on in the trees, back and forth and back and forth.

Nonetheless, let me tell you: whatever Katy did, she must have done eons ago, because the insect song is so old within me that I remember the song without having any memory of it, as if my first memories of hearing them didn't just come after I was born but somehow were born *with* me.

And you? I have to ask: Do you ever feel the same? Do you ever remember something from so deep down that it seems like something you were born into, that you were born knowing, before you even knew what knowing was?

I know, I know. Maybe I've had a touch of bourbon and should quit my rambling and tuck in for the night. Besides, it's raining harder now. It's time for a little shut-eye.

But you know the story now, don't you? You know this night won't end, not now, not for a long while to come. Because though we don't notice, at least not yet, the flooding has begun.

But no—that's not true, not exactly—the flood—it began far before us, didn't it? Maybe even before we were born. It was born in us and with us, going generations back when prospectors first opened up a vein of coal in these mountains, back when the first truck sputtered up one of these winding roads carrying a load heavy with timber.

And the katydids? Maybe they couldn't care about poor Katy and her jealous heart. Maybe, just maybe, instead of talking about her, they're debating on and on about us, trying to decide if what's about to happen is a crime, and if it is, if we're to blame.

Forgive me. Maybe you could care less about insects. Maybe you think that my saying all this nonsense about katydids is just another human projecting a human story on a creature doing their own thing for their own reasons. And you're right.

Nevertheless, do you hear them? What happens when you cock your head to the side and listen what this earth—and its waters—has to say? Or are you asleep by now like I was, tucked in my bed?

If so, just know it won't be long before you might hear a pounding on your door calling out, *Hello??!!?* just as I did, drawing out that knock-knock vowel of *o*.

No, what I mean to say is this: Are you listening? The waters are coming. Even if it's not raining where you are, even if you've been in a drought for years and years, even if you shower with a bucket at your feet to catch the soapy runoff and lug it outside just to keep a few struggling plants alive.

Forgive me if I say, *Hello??!!?* knocking on one room, then the next. There may be nothing any of us can do, but please, turn off your screens, open your door, ask the land if it will hold when the waters come. If not, plant tree after tree after tree; in years to come, they may just be all that's between a house of terrified people and the weather-terrorized land about to slide.

Hello??!!? I will say, again and again, for as long as I'm able, begging every person I can, telling them to rise. For mercy. For the earth loved more than any other earth. Rise, please, rise. You must wake now. We must rise.

notes

1. James Still, *River of Earth: A Novel* (1940; Lexington: University Press of Kentucky, 1978). Still's novel is written in dialect. I've seen several "translations," and the epigraph included here is my own rendering. The original publication reads: "Oh, my children, where air we going on this mighty river of earth, a-borning, begetting, and a-dying—the living and the dead riding the waters?"
2. James Still, "Wolfpen Creek," in *The Wolfpen Poems* (Berea, KY: Berea College Press, 1986).

When Water Becomes a Weapon: Fracking, Climate Change, and the Violation of Human Rights

Kathleen Dean Moore

*A*s evening comes on, my friends and I look over rolling hills of wind-silvered grass and a stock pond beside a windmill, slowly turning. Silhouetted against a livid sunset, three pump jacks tilt up, tilt down, up and down, ceaselessly, metronomically, silently; we are too far away from the fracking fields to hear them thud and squeal. The stock pond turns pink as the sunset fades, and by the time we finally hike back to our car, the pond floats a silver ladder of moonlight. Long into the night, the water holds the light.

But water can hold darkness too, and that is the subject of this exploration—how water can be made into a dark and dangerous thing. This will take us on a journey into the blackness deep underground, where blind water seeps and shushes through sand and silt. Here, in this darkness, is where the industry of hydraulic fracturing is turning water, an element essential to all living things, into a weapon against life.

Imagine that we can make ourselves small enough to follow a root into a crack in the rock, and down, and down, between grains of sand and through porous rock. It's dark down here—a prehistoric dark, a Pliocene dark, the dark of the past and of the future, the dark of dreams and crypts. Hard to say how far we have crept down through the tiny spaces. Fifty feet? Three hundred? But suddenly, water flows between every grain of sand. We have reached the top of the water table. Below us is an aquifer, water-soaked sand resting on an impermeable layer of clay.

On Earth, there is more fresh water down here in the silent rock than there is on the green and frenzied surface. In uncounted aquifers, fissures, underground rivers, and saturated sands, vast volumes of water silently, slowly move downhill under the pressure of gravity and the weight of tons of rock. Geologists call this groundwater "cryptic water" because it is mysterious and undeciphered.

How can I describe the water? It is dull black, of course, so far from the light. Maybe it smells vaguely of life, because there is life down here, a complex biosphere of bacteria and other microorganisms—maybe a greater biomass of life underground than on the surface of the Earth. The water is cold. Or sometimes it's hot. Maybe it tastes stale. I don't know—the water might have been down here for ten million years, or longer even than that, at home in this black, shivering, watery world unknown to us.

Some of this water may be "dead"—that's what the geologists call it when the water is imprisoned between impermeable layers, never to be part of the hydrological cycles. But most of the water is moving, flowing—if I can use that term for a process so slow—across a clay, maybe, or a glacial till, a giant sheet of water straining through sand, fracture systems, and fissures. Occasionally, water will find an opening and emerge from the darkness in a spring or a rancher's well, an oasis, or even a river—in a sudden splash of light, tinkling, twinkling, released, free as a fish.

And then it will embark on the next stage of its hydrological cycle in a cottonwood swale, or in a bison's flicking tail or the fly it flicks, in a quaking aspen leaf, in wild strawberries or spinal fluid, or in a glorious thunderhead blooming purple over the bison range. This is what gives life to Earth—this water of light, this water of darkness—and the movement from one to the other through long space and time. This is what creates the abundance of Earth, the singing of rivers and children, the paradise of plenty. This is how Earth grows beings who turn their faces to the night sky and sing praises. This is why Earth is not Mars.

A heavy drill with diamonds in its teeth grinds through the gravel and sandstone, down and down into the darkness, maybe a mile or more through dozens of geological layers and pockets of fresh water. Then it turns and drills horizontally, maybe another mile under the bison range. Who knows the sounds that vibrate so far underground or the smell of hot steel on sandstone? Who knows the sizzle when the bit touches water?

Roustabouts line the wellbore with concrete. Then into the wellbore, they pump fracking fluid or "slickwater" under pressure. Fracking fluid begins as fresh water. Oil companies draw the water from lakes and streams and often from the groundwater. How much water? It depends: somewhere between two and twenty million gallons of fresh water for each frack job.

In great slurry blenders, any of at least 1,021 chemicals can be mixed with the water to make the fracking fluid. It's hard to say what they are exactly, because the industry conceals much of that information. Trade secrets. But here are some of the chemicals: Benzene. Toluene. Ethylbenzene. Xylene. Arsenic. Cadmium. Formaldehyde. Hydrochloric acid. 2-Butoxyethanol. Ammonium chloride. Mercury. Glutaraldehyde. Other secret poisons to kill the microorganisms in the rocks before they can gum up the drill. These are chemicals customarily used to kill insects, clean toilet bowls, strip paint, polish brass, etch glass, preserve corpses, and commit murder. At least 157 of the fracking chemicals are reproductive or developmental toxins, causing birth defects, breast and prostate cancer, miscarriage, and other heartbreaks. The health effects of an additional 781 chemicals used have not been studied.

Now slippery, gelatinous, and entirely poison, the fracking fluid is forced down the wellbore under tremendous pressure. Twelve thousand pounds per square inch? Nine thousand? Numbers vary, but approximately the pressure a hand would feel if it were crushed by a steamroller. The fluid hits an opening in the concrete liner, explodes out with enough force to crack rock into shards and open long fissures. Silica sand is sent down to prop open the caverns, and oil and gas ooze or gusher out, beginning their complex transformation into money.

As poisoned water becomes the explosive weapon that smashes rock, this process is called "hydraulic fracturing." It might more properly be called "weaponizing water" against the very Earth.

Some of the poisoned water is forced through fissures in the rock, where it seeps into the underground aquifers, carrying radium that it absorbs from the rocks themselves. But between 18 percent and 80 percent of the used fracking fluid—what we used to call "water"—is brought to the surface. What happens to it then? Some of it is stored in open impoundments, which may or may not be lined to prevent seepage. Some of it is dumped into streams. Some of it is sprayed onto agricultural land. Some of it is evaporated from pits, the residue used to melt ice on highways.

But much of the fluid waste is injected back into the earth, where it finds its way along faults, through sands, eventually into the groundwater and some into well water, in an inevitable process called "migration," as if the toxins were birds or wildebeests. There are more than 480,000 underground waste injection wells in the United States alone; 30,000 of them force fracking fluid thousands of feet through water-bearing layers underground. No one knows how many wells are leaking. No one knows how much toxin finds its way into babies and breasts, into forests and agricultural land, into rivers and so into rice, where water becomes again a weapon, an agent of darkness and death.

The Dakota Access Pipeline carries "sweet" crude oil from the Bakken oil fields in North Dakota to oil terminals in Patoka, Illinois. In its 1,172 miles, it crosses hundreds of streams and burrows under twenty-two bodies of water, including Lake Oahe, an impoundment of the Missouri River that provides drinking water to the Standing Rock Sioux Reservation. Fearful of the good chance that the pipelines would leak into water that sustains their people, the Sioux rallied to stop the pipeline. These are the famous Water Protectors of Standing Rock. Hundreds of people came to help them stand their ground. Setting up encampments along the planned route of the pipeline, the people moved to block the progress of the great machines. The pipeline company, Energy Transfer Partners, brought in private security officers, the governor called out the National Guard, and local law enforcement officers moved in to clear the protesters.

Imagine the Sacred Stone Camp on a frigid night in midwinter. Searchlights flash through darkness that echoes with cries of "Water, not oil," "Water, not oil." Tear gas and the smoke of concussion grenades sting the night air. On one side of the Cannonball River is a phalanx of armored vehicles and police in full riot gear. Facing them: a crowd of Water Protectors, some wearing raincoats, but others protected only by plastic garbage bags and goggles. As shouts and rubber bullets zing across the river, the officers bring out their most powerful weapon. Water.

With water cannons as fierce as fire hoses, law enforcement officers blast the Water Protectors, knocking the people off their feet, tearing their clothes, and drenching them in ice water. People scream and run, trying to protect their faces from the force of the cannons. "We are cold. We are shaking. We are wet. We are in pain," one woman said, assaulted by the sacred element they were trying to protect—water turned into a weapon.

"Mni wichoni." "El agua es vida." In any language, water is life. But when water is made into death, we enter a sinister alternative moral universe where wrong is right, and profit is valued more highly than life itself. There is a breathtaking moral nastiness in wielding deliberately pressurized or poisoned water—naturally the source and sustainer of life—as a weapon against life. There was a time, and the time will come again, when this is morally unthinkable.

The wrongs begin as the oil corporation draws fresh water into its tanks. It is undeniably true that the life of every person on the planet depends on the 1 percent of Earth's water that is fresh and available. Of this limited supply, the US fracking industry uses an average of 105 billion gallons each year—as much as the water use of three million citizens of Chicago. Worse, this water is often seized from water sources essential to local people and extracted from some of the most arid and water-starved places on the planet. Because no technology exists to return fracking waste to potable water, this water becomes removed, maybe forever, from the hydrological cycle—a dangerous waste of the rare and wonderful gift of water.

Move on to the slurry tanks, where the symbol of innocence and purity, that agent of cleansing and renewal, is laced with the seeds of death. No one knows how the poisoned water spreads through the lacework of rock formations or human veins. Thus, in the 2018 judgment of the Permanent Peoples' Tribunal, a respected international opinion tribunal designed to shed light on human rights abuses of parties who lack access to justice, fracking practices constitute "deadly, large-scale experiments in poisoning humans and nonhumans that the fracking industry is currently conducting in violation of the Nuremberg Code."[1] The judgment is particularly damning: the nations of the world wrote the Nuremberg Code after World War II to forbid that any government, ever again, would experiment on human beings the way the Nazis did in the death camps.

Moreover, the wide-scale contamination of fresh water is a violation of the Universal Declaration of Human Rights, which guarantees that "everyone has the right to life, liberty and the security of person." These are not just words; these encode the moral consensus of the nations of the world, with none dissenting—an extraordinary agreement reached after World War II. The declaration sets hard ethical boundaries, minimal standards of human decency, recognizing that, as in its preamble , "disregard and contempt for human rights have resulted in barbarous acts which have outraged the conscience of mankind."[2]

Clean water is a necessary condition for the exercise of the guaranteed right to life. Thus, UN Resolution 64/292: "The General Assembly... recognizes the right to safe and clean drinking water... [as] a human right that is essential to the full enjoyment of life and all *other* human rights."[3] When fracking contaminates drinking water, it is an encroachment on this right. When fracking contaminates a river or stream that people depend on for drinking water, that is an encroachment on this right. When fracking fluid sickens an unborn child, a child, or an adult, that is a clear violation of the rights to life, liberty, and security of person.

Clear enough. But when people protest against the violation of their right to fresh water, they run up against the violation of other rights—the right to peaceably assemble and speak their minds. In fifteen states, soon to be twenty-two, it is a felony to "impede"—literally, to make forward progress more difficult—the operations of a pipeline or power plant. Even as fracking companies weaponize the water, they militarize law enforcement, arming local law officers with the surplus equipment of military forces and degrading the processes of democratic decision-making.

But as important as these violations of human rights are, it's when we go down with the drills into the seams that we encounter immorality even more grave, and that is the fracking industry's assault on the sanctity of water and the life-sustaining systems of Earth.

If there is anything on Earth that is sacred, it is water. *Sacred* means many things to many people. To me, it is the good English word that describes what is irreplaceable, beautiful, mysterious, powerful, essential, astonishing, and beyond human control or creation—sacred water, holder of light, holder of darkness, holder of all life. If water is sacred, then when fracking companies take it from Earth and from the people who depend on it, that is a sacrilege—*sacrilege*, the stealing of sacred things, from *sacra*, sacred, and *legere*, to steal. And destroying that water, wasting it, despoiling it, using it as an agent of destruction? That is a *profanity*, literally, *pro-*, outside of, *fanum*, the temple—taking lightly the attributes or acts of God. Water has a terrible power; when oil industries take it lightly, when they profane it, when they tease it with unthinking hubris, when they fail to show it proper respect or fear it fully, they create consequences of cosmic proportions.

Finally, we come to the truly world-destroying power unleashed when fracking industries transmogrify water into an explosive device. When fracking shatters ancient layers of rock, it releases the carbon that sank into prehistoric swamps and has slept there for two hundred million years, trapped in its underground crypts. Once freed by the explosive force of water, the carbon surges, snakelike, up the wellbore: crude oil and natural gas. After great amounts of money change hands, the carbon is burned. That releases carbon dioxide that traps enough heat in the atmosphere to irredeemably disrupt the systems that sustain life on Earth.

What systems? As we now know to our sorrow, climate warming caused by burning oil and gas disrupts the patterns of the wind, the force of the waves, the great currents in the seas, the reliable rivers of rain, the patterns of heating and cooling that allowed life to evolve in all its earthly sweetness and ferocity. Now, truly, the oil industry has unleashed watery weapons that lash out blindly, striking far more fiercely than a water cannon. Floods, hurricanes, rising sea levels, drought, saltwater intrusion: once water is made a weapon, it cannot be controlled. Climate chaos is the ultimate aggression, as the oil industry in so many ways enlists water as a foot soldier in its war against the world.

My friends and I found a strip motel not far from the fracking fields. Part of a man camp for the roustabouts, the beds smelled of cigarettes and hard use. Recoiling into the night, we sat under the flashing light of the motel's marquee. On the western horizon, methane flares glared off black clouds that rolled eastward until they erased the stars. The wind rose, the electricity blinked out, and rain began to fall. Big drops plonked on the dust, and suddenly the world was nothing but darkness, mud, and sage, as if we had been carried in our aluminum lawn chairs back into the mysterious eons when water created the world.

If we persevere, if we hold hard to what is right and name what is wrong, if we wrest control of water from extractive industry, our children may find a time when water can reclaim its innocence and rain remake the world.

notes

1. Thomas A. Kerns and Kathleen Dean Moore, eds., *Bearing Witness: The Human Rights Case against Fracking and Climate Change* (Corvallis: Oregon State University Press, 2021), 305. The full text of the Advisory Opinion of the Permanent Peoples' Tribunal on Fracking and Climate Change can be found at https://www.permanentpeoplestribunal.org/category/jurisprudence/?lang+en.
2. For the full text of the Universal Declaration of Human Rights, see the United Nations web page: https://www.un.org/en/about-us/universal-declaration-of-human-rights.
3. UN Resolution 64/292, https://digitallibrary.un.org/record/687002?ln=en#record-files-collapse-header

Water in My Veins: Reflections of a Coastal Indigenous Scholar

Clifford Gordon Atleo

When a driver starts a diesel engine, my visceral response is immediate. I am instantly brought back to the summer I spent in my youth on the western coast of Vancouver Island, British Columbia, working on my uncle's fishing boat. The droning rumble of the engine triggers powerful memories. I can see the morning fog that used to keep the air cool, even in the middle of summer. My body feels the motion of the waves, especially the big rolling waves that were calm and terrifying all at once. Even the smell from the diesel exhaust is oddly comforting.

Working for my uncle Harvey that summer was my first real job. *Sea Fury* was a forty-two-foot wooden troller fishing out of the Ahousaht village of Maaqtusiis. As a junior deckhand, my day began at about three-thirty in the morning. If we were already on the fishing grounds, we would drop our lines, and the seemingly endless cycle would begin: check the lines, clean the fish, ice the fish, check the lines, clean the fish, ice the fish. The work was not always intense, but it was steady. Over the course of eighteen hours, we would sneak in three meals, joke with one another, and constantly wonder whether we were going to get lucky with a big school of salmon. Perhaps my cousin David and I would wonder, but my uncle Harvey would brood in the wheelhouse with his coffee, looking out over the waters that our ancestors had fished for millennia. Of course, the equipment changed, and technology "improved," but salt water was in his veins. I believe it's in my veins, too.

Many coastal Indigenous people refer to themselves as "saltwater people."[1] Although I did not ultimately grow up to be a fisher like most of my ancestors, the ocean salt water and the fresh water of our rivers and lakes, and a life intimately connected to those environments, have factored heavily in my life and research as an Indigenous scholar. The work I do is one way I try to live in reciprocity with water, an effort to return its many gifts of life and livelihood.

I come from two coastal Indigenous nations and carry several Indigenous names. On my father's (Wickaninnish) side, I am Nuu-chah-nulth-aht, from the Ahousaht First Nation and the House of ƛaakišpiił (whale fat). I carry the name Čačim'mułnii, which roughly means "one who does things properly." I am Tsimshian on my mother's side. My mother's name is Gyemgm hup'i and my first Tsimshian name is Kam'ayaam, which has been passed down for many generations. It means "only imitating raven." We are from the House of Nishaywaaxs from the Kitselas First Nation located on the Skeena River. I also carry the title and name Sm'oogyit Niis Na'yaa.

Another important affiliation is that we are Gispwudwada, or from the Killerwhale Clan. Among the Tsimshian and neighboring Nisga'a and Gitxsan nations, all Killerwhales (and Fireweed Clan members) are considered close relations. These names and house affiliations situate me within a complex web of relations that is thousands of years old. They are reminders of who I am as well as of my responsibilities, not only to my human relations but also to my other-than-human kin and homelands and homewaters as well. This respect for interconnection and relations is common among diverse Indigenous worldviews. The Anishinaabe and Haudenosaunee scholar Vanessa Watts argues that everything in nature possesses agency and that we ought to think of ecosystems and all life within them as peoples and nations and not as things.[2]

I have been fortunate to learn and embody the belief that our lands and waters should be treated as relatives with respect. I currently live and teach on the lands of the xʷməθkwəy̓əm (Musqueam), Skwxwú7mesh (Squamish), and Səl̓ílwətaɬ (Tsleil-Waututh) Nations in Greater Vancouver. I do not live in my home territories, but I strive to be a good "guest" and remain connected to my home territories through family, food, and research.

To illustrate this, I want to share a little of my time living in the territories of the Mashantucket Pequot and Mohegan tribes. In 2014–2015, as a PhD candidate, I spent a year as a visiting scholar at Yale University. I lived in New Haven, Connecticut, with my wife, our two-year-old son, and our newborn daughter. This was my first time visiting the Eastern Seaboard and first time living in the United States, and it was an eye-opening experience. One of the first things that struck us was the huge disparity between rich and poor. New Haven epitomizes the inequality of neoliberal capitalism, as at various times it has been one of the poorest cities in United States, but Yale remains one of the wealthiest universities in the world. At the time, Yale had an endowment of about $26 billion plus $100 billion in assets. As a young family with two small children living off my scholarship income, we had to keep a strict budget, but we also felt tremendous privilege to be at such a prestigious school while also feeling disillusioned about the poverty that surrounded it. We did not know many people, and in retrospect, it was a very lonely time for me, writing my first draft of my dissertation, and for my wife, raising a toddler and newborn. One of the things that kept us connected to home was food from our families.

At one point, my father sent us two cases of canned salmon that were caught in Nuu-chah-nulth waters. I can't even imagine how much it must have cost him to ship to us, but we were grateful to have a small connection to the West Coast during that cold, snowy East Coast winter. We were able to nourish our minds and bodies with sockeye salmon from my homelands and homewaters. I recognized that those fish traveled a long way, at considerable cost, but they kept me connected to home. Even my dissertation writing felt more meaningful, connected to the salmon through physical nourishment and sense memory—a way for me to remember the water in my veins.

We also benefited from the generosity of my wife's family, who are from Curve Lake First Nation in Ontario. Her family sent us homemade maple syrup and traditionally harvested manoomin (wild rice). Consuming food from our traditional homelands kept us alive and allowed us to grow physically, mentally, and spiritually. I (Nuu-chah-nulth and Tsimshian) and my wife (Anishinaabe) are forever grateful for the food from our families and homelands that fed our bodies, minds, and spirits while we lived abroad. Pacific salmon begin life in the freshwater rivers and streams before venturing out into the saltwater of the ocean and swim great distances before returning four years later to spawn and begin the cycle of life again. Manoomin may be the bane of contemporary settlers in cottage country with their boats and Jet Skis, but wild rice has nourished my wife's people for countless generations, and we remain grateful that her family carries on those traditions. Sadly, many of these vital connections to our territories remain threatened. The protection of Indigenous waters and homelands has been central to my research as a graduate student and now as a professor.

Yet honoring and protecting these waters is not always straightforward. The year 2002 was an important one for me as an Ahousaht and an Indigenous scholar, as several threats to our homelands and homewaters reared their ugly heads. At the time, I was working for the Nuu-chah-nulth Tribal Council (NTC). In the spring, a newspaper headline read "Ahousaht Declares War against Fish Farms."[3] The Ahousaht, particularly the Ha'wiih (hereditary chiefs), were opposed to the expansion of fish farms in Clayoquot Sound, especially because an estimated ten thousand Atlantic salmon had escaped from one of the open-net pens. There are many issues with fish farms, but major concerns include the colonization of local streams by Atlantic salmon and the threat of sea lice to local Pacific salmon. By the autumn of the same year, the Ha'wiih had signed an impact benefit agreement (IBA) with the fish-farm company that allowed for the farm's expansion. IBAs represent a new era of Indigenous-settler engagements around extractive industries in Canada. IBAs are almost always confidential agreements, negotiated with an industry proponent and an Indigenous community, that bring the infringement of Indigenous rights within the realm of contract law. Partly because settler governments move slowly, corporations have accepted these new costs of doing business and often preemptively engage Indigenous nations on resource projects to avoid conflict.[4] I was alarmed by the Ha'wiihs seemingly quick about-face on the issue of fish farms, and I wanted to understand why they made their decision. Another event in 2002 added important context to my understanding.

Five Nuu-chah-nulth nations (Ahousaht, Ehattesaht, Hesquiaht, Mowachaht/Muchalaht, and Tla-o-qui-aht) launched a case in provincial court claiming an Aboriginal right to participate in commercial fisheries. While Nuu-chah-nulth people historically obtained as much as 80 percent of our sustenance from whales that we hunted and salvaged, commercial overharvesting by non-Indigenous whalers had led to a cessation of traditional whaling by the end of the nineteenth century.[5] For the majority of the twentieth century, Nuu-chah-nulth-aht heavily participated in Pacific coastal commercial fisheries, mainly for salmon, but also for cod, halibut, and other species. I have described this elsewhere as an adaptive livelihood, as it allowed Nuu-chah-nulth-aht to remain connected to a life at sea, albeit one regulated by settler governments.[6]

At the peak of Nuu-chah-nulth participation in these commercial fisheries, we owned and operated approximately two hundred large fishing vessels. By the time the *Ahousaht* fishing case was launched in 2002, we had only six boats left. Within the span of a single generation, we found ourselves moving from a situation in which practically every Nuu-chah-nulth citizen was directly related to someone participating in these fisheries to a time that nearly no one was. In many ways, we had ceased to live as salt-water people. By the turn of the century, Nuu-chah-nulth-aht had been nearly starved into submission. And although the Supreme Court of Canada upheld the lower court's ruling that the five Nuu-chah-nulth nations did in fact possess commercial fishing rights in 2014, negotiations on the implementation of the ruling remain on-going with the federal government. As opportunities to maintain livelihoods, fishing wild salmon has greatly declined and remains precarious; fish farms have only grown internationally despite the controversies and criticisms of them.

Citing the Marxist philosopher Rosa Luxemburg, the Sámi scholar Rauna Kuokkanen writes: "Only by destroying their capacity to subsist are people brought under the complete control of capital. Coercion is needed to destroy not only the capacity to subsist but also a people's economic and political autonomy."[7] By the end of the twentieth century, Nuu-chah-nulth-aht had been driven out of our livelihoods at sea. Prior to its effects on commercial fishing, non-Indigenous commercial extraction had also greatly affected pelagic seal hunting, the maritime fur trade, and whaling.[8] With few options left, Nuu-chah-nulth-aht have found themselves between the proverbial rock and a hard place. Although it was not readily apparent to me at the time, in retrospect, I better understand the difficult decisions that our leaders had to make regarding farmed fish.

The early 2000s also marked the incursion of mining companies into our territories, and our loss of traditional and adaptive livelihoods forced us to be more receptive to their overtures. The first company to approach the Ahousaht in 2000 about a potential copper mine in Clayoquot Sound was Doublestar Resources. It held a tenure for Chitaapii (Cat Face Mountain), and initially the Ahousaht were resistant. Tyee Ha'wilth (head hereditary chief) Maquinna stated that Chitaapii was sacred and that its destruction would never be worth it.[9] The tenure was later acquired by Selkirk Minerals and ultimately Imperial Metals. Despite early opposition to the mining companies, economic conditions worsened for the Ahousaht, and in 2008 they signed a memorandum of understanding with the mining company to allow exploratory drilling. As in the case of the fish farms, Ahousaht citizens were ambivalent, and the decisions have been controversial. At the same time, socioeconomic concerns greatly affected the decision-making process. In fact, at one point, the chairman of Imperial Metals claimed that his company was committed to the mine project as a social obligation to address the poverty of the Ahousaht.[10]

Ultimately, the Ahousaht rejected the mine proposal in their 2017 Land Use Vision, after extensive community-wide consultation, but I fear that the growing demand for rare minerals and metals to power the postcarbon economy will make Chitaapii a mining priority again. This is not simply an issue of mineral extraction for the new economy; it is also a matter of water quality and our responsibilities to water. One of the primary concerns about mining is the tailings produced from extraction and the subsequent contamination of water. In this case, Ahousaht citizens were concerned about contaminating our fresh and salt water. The impacts of that would be felt not only by people but also by our other-than-human kin as well. The risk was high, but acute poverty demanded that we consider it. The continuous threatening of Indigenous lands and waterways becomes a version of environmental racism, a concept originally coined by Benjamin Chavis.[11]

Environmental racism, when identified, can also call forth a collective response. It was a threat to Indigenous waterways that led to one of Canada's largest sustained Indigenous social movements: Idle No More (#IdleNoMore on social media). Although Indigenous resistance movements have been around since before Canada was a country, Idle No More was significant in that it resonated with so many people, young and old, Indigenous and non-Indigenous, radical people and people who might never consider themselves radical. In late 2012, the Conservative government under Prime Minister Stephen Harper introduced Bill C-45, a massive omnibus bill that proposed industry-favoring changes to the Indian Act, the Fisheries Act, the Canadian Environmental Assessment Act, and the Navigable Water Act. Four Saskatchewan women (three Indigenous and one non-Indigenous)—Sylvia McAdam, Jessica Gordon, Sheelah McLean, and Nina Wilson—sought to educate the public about the government's intention to remove protections for Indigenous territories, especially water.[12] Idle No More grew as a grassroots movement that was explicitly nonviolent and featured events with singing, drumming, and round dances across Canada. Solidarity demonstrations took place all over the world, including in the United States, Europe, Aotearoa (New Zealand), and Africa.[13] My wife and I and our infant son participated in two demonstrations that winter—one in Nogojiwanong (Peterborough, Ontario) and one in W̱SÁNEĆ territory (near Victoria, BC). Although the movement had little impact on changing legislation, it did raise the consciousness of a new generation of Indigenous activists and encourage many new non-Indigenous allies. The struggle continues to gain recognition for the sacredness of water as our source of life and livelihood and as a complex set of living relations.

I want to share some final thoughts on a Nuu-chah-nulth ritual intimately linked with water: *uusimč*. *Uusimč* is often understood as ritual cold-water bathing or cleansing that is undertaken only at certain times in the lunar cycle in the winter months. Nuu-chah-nulth-aht engage in *uusimč* (which often includes prayer and fasting) when they wish to acquire knowledge from the spiritual realm or prepare for a great task (like a whale hunt). Because it is normally a secretive ritual, and each family has their own specific methods, I will not describe it in detail. I will share that my uncle Umeek calls it a "careful seeking in a fearsome environment."[14]

My father was the first to teach me how to uusimč. The experience is exhilarating and tremendously humbling. Umeek emphasizes that a humble approach is key to success and that egotistical orientations from a Nuu-chah-nulth worldview are often thought to be unsuccessful.[15] I have been cleansed by icy cold coastal waters and the hot steam of the sweat lodge of my interior Indigenous relatives, and the one thing they have in common is that they demand my humility. These rituals and ceremonies remind us of humanity's place as the younger sibling of creation.[16]

Honestly, it has been too long since I uusimč'd. I am overdue, and I think others are too. The earth and all of creation are suffering because of our collective arrogance and greed. I conclude with the Nuu-chah-nulth principle of *hiišuukiš čawaak* (everything is one). We are all connected, and it is incumbent on us to engage in our ceremonies that remind us of our place in creation and our responsibilities to uu-ałuk (take care of) one another.

I may not work on a fishing boat, as I did during the memorable summer with Uncle Harvey, but I carry those memories within me and, as I do, my responsibilities to our homewaters as a person deeply connected to water-dependent Indigenous communities. I continue to try to care for the water as it continues to take care of all of us so that we may live with respect and reciprocity as the younger siblings of creation.

notes

1. Clifford Gordon Atleo, "Change and Continuity in the Political Economy of the Ahousaht" (PhD diss., University of Alberta, 2018).
2. Vanessa Watts, "Indigenous Place-Thought & Agency amongst Humans and Non-humans (First Woman and Sky Woman Go on a European World Tour!)," *Decolonization: Indigeneity, Education and Society* 2, no. 1 (2013): 20–34.
3. David Wiwchar, "Ahousaht Declares War against Fish Farms," *Raven's Eye* 5, no. 9 (2002), https://ammsa.com/publications/ravens-eye/ahousaht-declares-war-against-fish-farms.
4. Clifford Gordon Atleo and Jonathan Martin Boron, "Extractive Settler Colonialism: Navigating Extractive Bargains on Indigenous Territories in Canada," in *Extractive Bargains: States, Resources and the Elusive Search for Consensus*, ed. Paul Bowles and Nathan Andrews (Berlin: Springer Nature, 2023).
5. Charlotte Coté, *Spirits of Our Whaling Ancestors Revitalizing Makah & Nuu-chah-nulth Traditions* (Seattle: University of Washington Press, 2010).
6. Atleo, "Change and Continuity."
7. Rauna Kuokkanen, "Indigenous Economies, Theories of Subsistence, and Women: Exploring the Social Economy Model of Indigenous Governance," *American Indian Quarterly* 35, no. 2 (Spring 2011): 223.
8. Atleo, "Change and Continuity."
9. Clifford Gordon Atleo, "Nuu-chah-nulth Economic Development and the Changing

Nature of Our Relationships within the Ha'hoolthlii of our Ha'wiih" (MA thesis, University of Victoria, 2010).
10. Atleo, "Change and Continuity."
11. Robert D Bullard, *Dumping in Dixie* (New York: Taylor and Francis, 2018).
12. The Kino-nda-niimi Collective, *The Winter We Danced: Voices from the Past, the Future, and the Idle No More Movement* (Winnipeg, MB: ARP Books, 2014).
13. Tim Groves, "#IdleNoMore Events in 2012: Events Spreading across Canada and the World," The Media Co-op, December 17, 2012, http://www.mediacoop.ca/story/idle-no-more-map-events-spreading-across-canada-an/15320.
14. E. Richard Atleo (Umeek), *Tsawalk: A Nuu-chah-nulth Worldview* (Vancouver: University of British Columbia Press, 2004), 84.
15. Atleo (Umeek), *Tsawalk*.
16. Shane Edwards, "Titiro Whakamuri Kia Marama Ai Te Wao Nei: Whakapapa Epistemologies and Maniapoto Māori Cultural Identities" (PhD diss., Massey University, 2009); Cash Ahenakew, Vanessa de Oliveira Andreotti, Garrick Cooper, and Hemi Hireme, "Beyond Epistemic Provincialism: De-provincializing Indigenous Resistance," *AlterNative* 10, no. 3 (2014): 216–31; Ingrid Leman Stefanovic and Clifford Atleo, "Valuing Water," in *Ethical Water Stewardship*, ed. Ingrid Leman Stefanovic and Zafar Adeel (Cham, Switzerland: Springer, 2021).

Spirit Walking in the Tundra

Joy Harjo

All the way to Nome, I trace the shadow of the plane as it walks
Over turquoise lakes made by late spring breakup
Of the Bering Sea.
The plane is so heavy with cargo load it vibrates our bones.
Like the pressure made by light cracking ice.

Below I see pockets of marrow where seabirds nest.
Mothers are so protective they will dive humans.

I walk from the tarmac and am met by an old friend.
We drive to the launching place
And see walrus hunters set out toward the sea.
We swing to the summer camps where seal hangs on drying frames.
She takes me home.
I watch her son play video games on break from the university.

This is what it feels like, says her son, as we walk up tundra,
Toward a herd of musk ox, *when you spirit walk.*
There is a shaking, and then you are in mystery.

Little purple flowers come up from the permafrost.
A newborn musk ox staggers around its mother's legs.

I smell the approach of someone with clean thoughts.
She is wearing designs like flowers, and a fur of ice.
She carries a basket and digging implements.
Her smell is sweet like blossoms coming up through the snow.
The spirit of the tundra stands with us, and we collect
    sunlight together.
We are refreshed by small winds.

We do not need history in books to tell us who we are
Or where we come from, I remind him.
Up here, we are near the opening in the Earth's head, the place
    where the spirit leaves and returns.
Up here, the edge between life and death is thinner than
    dried animal bladder.

(for Anuqsraaq and Qituvituaq)
Nome, Alaska, 2011

In the Land of the Five Seasons

Martin Lee Mueller

The basic structure of regilaul song: a verse is sung by the
leader and repeated by chorus, usually joining on last syllable
of the verse.

—Taive Särg, "Does Melodic Accent Shape the Melody
Contour in Estonian Folks Songs?"

He had been kissed by the ghost of the raised rainy bog. If he had been able to choose, he would not have come here in the first place. He stood before the crowd and spoke, and he appeared so reluctant, as if he wanted nothing more than to become one more shadow among the many pine shadows. The low-hanging evening sun flooded the clearing with a golden glow. Some folks leaned against haggard tree stems. Others sat on sheepskins and felted cushions or used their rucksacks as pillows. Moss slurped and squelched between our naked toes. The air smelled foul. Someone had laid out homemade cakes on a patchwork rug beside large jars of lemonade with berries and mint leaves. There was coffee and tea, and those who wanted could buy a dream catcher formed by hand from twigs, feathers, shells, and colored shards. Some of the youngest children slept in the arms of grandmothers. Someone had spanned a hammock between two pines.

All eyes were on him.

He had been addressing the audience for some time now, in that wary murmur that concealed more than it revealed. I had been observing him. I was not entirely sure whether he was a human at all. His hair fell across his face in dark waves, curtaining the forehead, eyes, nose, mouth, chin. He hid behind a veil of timidity and evening shadow. And yet there he stood, set to sing to hundreds of us. He said outright that he had never done anything like this. He said he also had no choice.

I remember you. Your legs still moved with the rhythm of the last *regilaul*, or runic, song. It was way past your bedtime, but what did that matter? It was one of those Nordic summer evenings that would not exhaust itself in darkness. Nothing kept you on your mother's lap. You had to stand, you had to move! Your body bobbed back and forth in ever-identical repetitions. The ground sagged and gurgled with each of your steps. You looked at your feet. You burst out laughing. The pine clearing was a spongy island in what seemed like an endlessly extended moorland made up of nothing but mosses and grasses and dwarf trees, a wide-open expanse stretching to the horizon in every direction. Here and there, the bog expanse was interrupted by mirrorlike ponds. In each one of them the sky touched itself. Every pond seemed like a hole stamped right out of the landscape, a nothingness flooded with water. Every pond was a harbinger ten thousand years old, heralding from a time when this whole landscape here had not been land at all but rather the lake bottom of the giant Baltic Ice Lake. Yet a few thousand years further back, the glaciers had retreated, and the lake had formed from meltwater. Eventually the lake retreated here too, and the land began to rise. Some lowlands held on to the water even then. And that's where it all began.

Perhaps it began with the cotton grass. Cotton grass seeds can wander through entire sweeps of land, free and unbounded. Swathed in fluffy woolen tufts, the seeds can travel on a breath of air or sail across the waters. They abandon themselves to the touch of waters and winds. In that touch, they arrive fully within themselves. Their bodies are living expressions of their trust in being carried by Being. They are certainty expressed as form, meaning that life finds itself only in radical abandonment.

The seeds of cotton grass mediate a liminal space of possibility between seclusion and unfolding. Seed pods encase the dawning of primordial wills. Sun and rains will tickle forth an urge inside each seed, to open itself to the sky and the earth. Once the will has started sprouting, it flows with increasing vigor toward the form of the plant. Cotton grass can settle into sites where there was nothing before. It knows that life means constantly reclaiming and sustaining the desire to go on living. Cotton grass has a strategy. The plant creates itself in its body structure, its habits, and its entire life cycle as one coherent and self-contained energy flow. Its leaves reach toward the light in tight alignment, careful not to sprawl too far from the center. Meanwhile, the grass will root intensely through the upper soil layers in every direction, preparing, preparing. For if... when... some of the leaves wilt, then fall off, they will fall upon a receptive ground which *is* soil and *is* body. Cotton grass interprets its inevitable overground dying as the enduring well to which it can keep returning, imagining itself anew in the next creative swell, rechanneling that energy she received from her own dying body toward new leaves. Cotton grass unfolds its life as a periodic circular flow of its own body. Flow itself becomes a seam of opportunity, a niche on the razor edge between impossibility and creative emergence. This is how the grass's own dying rumors of future renaissance. If there is such a thing as an essence of the grass, it is inseparable from its self-composed, imaginative flow. The grass *is* imagination *is* self-becoming flow *is* rebellion against death *is* ever-renewing will *is* world making from within the fertile abyss of nothingness.

Rainer Maria Rilke comes to mind. Rilke draws the image of a wretched panther behind bars, the miserable predator who, for a century, has been doing nothing but "pac[ing] round and round in cramped circles,... like a dance of power around a center, in which a potent will stands paralyzed."[1] Cotton grass gives a variation of

Rilke's theme, of a will that gravitates toward its own center. Cotton grass seems to suggest that the theme is a universal quality of the phenomenon of aliveness. The important difference is that in the case of the grass, the will is anything but cramped or paralyzed. It emerges primordially from flowing around its own center. Paradoxically, it is when the body flows in self-referentiality that it gains the freedom to keep opening itself to the world.

And yet the grass will eventually perish.

When it does, the impulse of its life will not ebb away completely. The gravity of its self-referential will may have dissipated, but some of the impulse moves forward. At the hour of death, the impulse reimagines itself, not as the impulse of a body, but of the land itself. When cotton grass dies, the thick leaf blades disintegrate into heaps of fibrous tufts. These are in part woody, for they contain lignin. The fibers sink into the oxygen-poor waters around the grass, where they will not degenerate entirely. As time goes on, what emerges is peat fiber, a spongy, soaked, nutrient-rich, anaerobic organic structure that will lift an entire landscape above its surrounding terrain. A raised rainy bog is born.

Cotton grass is the promise of metamorphosis expressing itself as a body. It mixes water and minerals from below with sunlight and carbon dioxide from above, mixing earth and sky so thoroughly that aliveness flows from their intermingling. Cotton grass composes mere potentiality into coordinated self-becoming. With every new life cycle, cotton grass yearns toward the sun and the earth, rises up against scarcity, defies gravity, dares death, arranges bodies from rain and wind and rays of light, and as the centuries pass, and as the centuries become millennia, its urging for life raises the bog higher and higher, letting a landscape become the expression of an unquenchable thirst for more life. Much of the bog's substance derives from carbon atoms and oxygen atoms filtered out of the breath-ocean by bog plants. Bogs are intermediaries between land, water, and air. They unify and channel uncounted circulating wills toward one larger circulation, that between the lithosphere, hydrosphere, and atmosphere. When they do, more and more carbon imagines itself as plant body; plant bodies imagine themselves as peat; peat imagines itself as bog; the bog imagines itself across the millennia as cooler climate, for as time goes, the bog pumps enormous amounts of greenhouse gas from the atmosphere. While it does, the self-becoming landscape continues to imagine itself as body—sundew, elf ring, the cry of cranes, the flight of dragonfly, wolf howl, salamander dreaming, rain song, she-bear's winter's sleep, sun dance, the laughter of a child.

Somewhere in between those layers of yearning—somewhere among this circulating self-realization-through-others—a world-spanning sphere of aliveness imagines itself into actuality. The biosphere may in that sense be just this: the playful and radically open surrender of bodies to landscapes and landscapes to bodies. The biosphere is the will to live, manifested and expressed in body cycles, life cycles, place cycles, regional cycles, larger cycles, as that will ever fathoms its own depths and constitutes itself as a sphere. Aliveness, it seems, is a cosmogenic phenomenon, a spherical dance of power around the burning center of our world in which potent wills behold themselves and see: It is as it is. It is good that way.

Above it all, the sun rises and sets, rises and sets. Dreaming itself, we may speculate, into the radiance of countless bodies as they dawn into selfhood.

By now, the nameless singer had started to sing. Whenever he had sung a phrase, several hundred repeated the phrase, grandparents and young parents, and the few children who had not yet drifted across into sleep, lulled in by the ever-repetitive pace of the *regilaul* song. The low evening sun pierced through the pines from the northwest. It almost seemed as if it encircled the evening in a cage of shadow-bars. We all sang and sang. We circled around one another. From within our midst, a story started to step forth. The story insisted on becoming voice, word, melody, cadence. The story searched for ways to manifest itself as will, as wish, as the power to sing itself into actuality.

Sooma, the land of bogs, is the land not of four but five seasons. There is spring, summer, autumn, and winter—and there is the season of the floods. Water masses flow from the Sakala Uplands and into the rivers that cut through this land of bogs. When they do, they overwhelm the rivers so that the water floods the swamp forests and grassy plains and, yes, the roads. Within days, the water level can rise by three or four meters, turning the network of raised rainy bogs into an archipelago of spongy islands in a seasonal ocean. The people who live there have kept alive the ancient craft of dugout canoes, allowing them to glide silently through their water world. It is a world they share with so many others. Tundra swan, golden eagle, whimbrel, plover, merlin, northern willow ptarmigan, harrier, pike, moose, wild boar, beaver, lynx. In each of them, the world intones itself in intensified inwardness. All are kissed by the ghost of the raised rainy bog. In their thousand-fold manifestations, the bog ever awakens to itself.

We sang and sang. I sat there on our makeshift camp of skins and blankets and observed you. You moved freely among the crowd, far from us, moving rhythmically, wafting weightlessly as in a dream. You were taken by a freedom you seldom showed in crowds. At two, you were a sensitive little animal, subtly alert, empathic, observant. Crowds often caused you to turn inward, withdraw a little. From within your inner calm, you seemed to observe more keenly: Where am I? Who are the others? Who am I in relation to them? In crowds you often narrowed your moving range, circulating closer to us, searching, if possible, for direct body contact. A fingertip, a pant leg, a belt, the tail of a jacket. You wouldn't need much. In our touch, you'd find the courage to be closer to yourself inside this large world. Except that tonight, you did not need our touch. You were already being touched. You were touched by a hundred voices. You were bedded in a commonwealth of shared aliveness. You were held by mutuality. In your surrender to the voices, you found a way to yourself. You were toddler, and you were tufted woolly cotton grass, were dance, were beam of light.

Perhaps the meaning of the raised rainy bogs lies in the way in which it fluidifies thinking itself. The languages I feel confident navigating—German, Norwegian, English—are all steeped in centuries of the Occidental intellectual canon. They are languages of things. They perform grammars of discrete stuff. They place us inside a world of already-articulated objects. *This tree. That rock wall. Those mountains. That river. This bog. This child.* The grammatical structure, with its emphasis on nouns, already assumes a static world. It principally discourages a curiosity about the ways in which trees, rock walls, mountains, and rivers *inter-are*. But that is neither an obvious nor an inevitable way of situating ourselves inside the phenomenon of aliveness through speech. There exist languages composed around "grammars of animacy," as Robin Wall Kimmerer shows in *Braiding Sweetgrass*. Kimmerer gives the example of her Native Ojibwe language, a language not of nouns as much as of verbs: Not things, gestures. Not stuff, movements. Not discrete entities, interactions. Not objects, unfolding relationships. To think: the tree may rather be encountered as *a treeing* (and *being treed*). The rock wall, as *rock walling* (and *being rock walled*). Mountains, as *mountaining* (and *being mountained*). The raised rainy bog, as *raised rainy bogging*. The river, as *rivering*. The child, as *childing*. To think that every spoken gesture can be a subtle invitation to bear witness to patterns of mutual touch! To think that speaking may be perceived as a standing invitation to keep calibrating the human condition inside vaster and more dynamic flows of relationship. To think that every "thing" may be a thinging, a uniquely embodied gesture. To think that every utterance may be perceived, not as a way of stepping away or outside the phenomenal world, but rather as a movement with, inside, and through a living world in constant flux. This makes all the difference. Grammars of animacy are grammars of relationship. They incline speakers to think *with* the world, rather than about it. If I see a *treeing*, I will be inclined to let my thinking follow the movements near and far: What are the exact, unique patterns of *that* treeing (rather than this one here)? How is the treeing expressing movements of water (upward, downward, outward, inward)? In what ways is the treeing letting breath pass through and transforming it on the way? How is it passing through the seasons (rhythmically!)? How is it passing across the years? How are its very different, very relaxed movements stretching my nervous animal imagination into a life design so entirely alien, yet not entirely apart? How is the treeing communicating and interplaying with other nearby treeings, fungusings, buggings, birdings, deerings, badgerings, boarings? How are their enlivened gestures, their sentient bodies breathing, receiving the gestures of the treeing, only to transform and condense them into impulses of their own unique life designs? How do the multitudes of gestures and transformations flow in and out of one another, becoming body, becoming nutrient flow, becoming voice, becoming inwardness, becoming *oikos*?

Likewise, if I stand face-to-face with a childing, I will be inclined to let my thinking follow the movements near and far: What are the exact patterns of *that* childing (rather than this one here)? How is the childing expressing movements of water, movements of breath, the passage of the seasons, or the passage of solar years, as all those flows spring to life and irradiate themselves inside the childing's flesh? How are this childing's familiar yet never fully fathomable movements stretching my imagination into a life design at once intimately known and too vast to ever be fully knowable? How does the childing communicate and interplay with other nearby bodyings, breathings, flowerings, rockings, insectings, riverings? How are their enlivened gestures, their sentient bodies breathing, receiving the gestures of the childing, only to transform and condense them into impulses of their own unique life designs? Further: What would it be like to live in such a world(ing) of mutual touch and interpenetration as a matter of basic speech habit? What would it be like to live inside the atmosphere of such an imagination, personally, collectively, across the span of a lifetime, across the span of a speech community's far longer unfolding? What would it be like to reorient ourselves inside such speech habitats? And what would it be like to practice this way of seeing with everything(ing) we encounter, collectively, over time, crafting cultures of animacy? Cultures of animacy are cultures fluent in the delicate art of navigating the world as self-transformative flow. This, we now understand, includes not only becoming fluent in biochemical, biological, geomorphic, or ecological animacies, but becoming fluent also in the ways we as humans may participate wisely with such more-than-human animacies. Fluency in enlivened realism integrates and enlarges empirical realism proper. Enlivened realism is more than a conceptual recalibration; it is more than a different way of speaking. It is a commitment toward the practical wisdom of becoming human, here inside this self-becoming biosphere.

All those movements of the imagination: encouraged by fluidifying—read: reanimating—thinking even at the simple level of grammar. Alerting us to the importance of speaking and listening with care. Drawing us deeper into relationship. Helping us, still, to arrive inside this self-transformative world. Not to crash-land, screaming. But to land with our feet more firmly on the rain-soaked, soggy ground that sustains us. To lend the composed breathing that we call human voice to the honoring, celebration, and continuity of the commonwealth of shared breathing that is the air.

On and on the singer who really wasn't a singer sang. His singing paced forward and through his *regilaul* song, stanza by stanza, leading us all, followed by us all. His dark curls still hung in front of his face. His eyes were withdrawn into a place somewhere near-complete inexpressiveness. Something in him was incensed. Here was no performer leading us. Here was someone chased. It was as if he were being whipped by unseen agencies.

Surely, this fluid way of thinking may sound strange from within the received thought-habits of an enlightened mind. That strangeness need not be a problem. It can be a promise. It can be a transformative edge. It can help prepare us for the possibility that we may experience a liberating estrangement in relation to our habitual grammars of discrete stuff. It can help prepare us for the possibility that even Occidental languages, if carefully spoken, can reanimate thinking far more than we may suspect. It can prepare us for the distinct possibility that we can rethink human agency, human voice, human transformations, and human commitments inside more-than-human agencies, voices, transformations, and commitments. It can prepare us for the possibility that we can yet become more fully human to the degree that we step more fully inside the phenomenon of aliveness.

The metamorphosis of the grass does not end with the swelling of the rainy bog. Allow for the agency of high pressures, high temperatures, and a complete lack of oxygen, and expand your view from thousands of years to millions, and the circles widen ever more. Peat becomes brown coal, then black coal, then anthracite, and ultimately, graphite. Perhaps—maybe—here is indeed an end to the circling, for at this point, ancient atmospheres have shape-shifted through the agency of the bog into pure carbon crystals. One can hardly imagine a more stable and more durable storage of greenhouse gases. With the alchemy of greenhouse gases into crystals, life creates for itself a huge tidal wave, a channel through which heat gets diverted across deep time, from the most volatile and treacherous state into the most solid and least potent state. Like singing, atmospheric coolness is a form of commonwealth. It takes all of life to help regulate the planet's own cool conditions, even as the sun grows ever hotter across the expanses of deep time.

Aliveness shows itself as cosmogenic phenomenon that flows with the whole earth and composes the fluid flesh of the real more deeply within itself. Aliveness cannot be reduced to bodies or to populations, species, or even webs of species. Aliveness is the geo-storical, storied, evolved, spherical ability of this planet to gather itself around the world-making force of countless wills to live and to imagine itself in new shapes, new bodies, new patterns. In the phenomenon of aliveness, the cosmos has infected itself with a nearly inextinguishable compulsion to encounter itself ever more deeply in inwardness.

He sang in his monotonous ductus. We answered. No variations to the beat. In singing, he wandered through landscapes scarred by drainage ditches. The bogs got drier. Complex ecological processes tipped over. I noticed how a newfound urgency took possession of his voice. The voice became pushier. Raspier.

Bogs are so successful in converting carbon into organic substances that they store more carbon than all other types of vegetation combined. In Europe alone, bogs sequester five times more greenhouse gases than forests. But these processes can tip. Most

European bogs are drier today than they have been for thousands of years. More dryness equals greater vulnerability. Every bog has its tipping point at which the dynamics of carbon sequestration become reversed. The landscape begins to emit carbon. It is the beginning of a monumental breathing out, one giant word spoken on the outbreath.

Heating.

The politically correct version of the story goes something like this: Estonia, your mother's homeland, considers peat its second most attractive resource after oil shale. Estonia may be tiny in size compared to many other European countries, but it ranks in the top ten of peat-rich countries in the world. More than two billion tons of peat are stored across one and a half thousand bogs, averaging a depth of four to five meters. One-quarter of it all is written off for industrial mining. The majority of it gets transported to Central Europe's garden centers.

Note how quickly the language of empirical realism slips into the abstract, the quantifiable, the global. Note how cultural expressions sculpted on such thinking tend to become insensitive to anything but the crudest expressions of agencies.

Where does the commonwealth of our songs begin? When and where does blood begin to pulsate? Where do bodies form into wills, and how do wills gather themselves around the creative potency of speech?

Step by step, he wandered through drainage ditches, physical manifestations of Enlightenment-style water-thoughts that had transformed living landscapes into graveyards. For years he had been wandering the ditches, gauging the depth of ignorance which had blood let his land. With every step, his feet touched millennia, self-composed into life. With every step, the ages trickled into his imagination, whispering images into his ear, regenerating his sense of the possible. With every step, intuitions emerged within him and condensed into emotions, which needed to move forward as shared rhythm. With every step, fractals of a vision for regeneration whirled up into his body-mind, like autumn leaves, becoming shared presence, clear-sightedness, farsightedness, clairaudience, voice.

Reverse the processes of peat harvest.

Refill the drainage ditches.

Stop the bog bleeding.

Let landscapes swell again.

Start again the genesis of raised rainy bogs.

Create the impulses the landscape needs to imagine itself once more as it had done for thousands of years and could still do for thousands of years to come.

Expand and recalibrate the imagination away from profit-driven, self-referential thinking toward speaking life as commonwealth, as ever forming and firming.

Help the land take a giant carbon in-breath again.

Even if you have nothing to give but the most introvert, timid, uncertain out-breath: give it. Every word spoken on behalf of aliveness is a gift in the proper sense. Every human voice pooled toward aliveness can help the land speak a different kind of word again, a word articulated in a speech so very unhurried and so very leisurely it would make Entish sound like rap.

Cooling.

What are proper perspectives to speak the phenomenon of aliveness succinctly and realistically, here inside the rotating Orion arm of the Milky Way Galaxy inside whirling breathspheres inside places inside human bodies? What is insight, if not the cumulative ability to sing praise to all these layered depths?

The singer who was no singer lived far off the charts of human attention, at the margins of his community, a deer who could not bear being stared at too long. But the bog insisted on letting voices rise inside him. The voices had no patience with his introversion. They left him no choice but to stand up and sing. The bog shaped his voice into her own in order to self-transform into a music that would resonate with human hearts and minds. He wandered the drainage ditches season after season, year after year. The pace of his feet became the pace of the story he would chant. Phrase by phrase. Stanza by stanza. Song was commonbreath turned to voice. Commonbreath passed between everything and everyone. Cotton grass. Dwarf pine. Black alder. Black stork. Bog moss. Human child. He sang because he had been shown: to live is to be oriented toward shared flow.

See, and you shall know.
We shall live yet again.
Do what you must.

Look, dear child: You too are breathing.

notes

1. My own translation from Rilke's "The Panther."

The Many Faces of Water

Mark Riegner

If there is magic on this planet, it is contained in water.
—Loren Eiseley, *The Immense Journey*

When I first stepped into John Wilkes's sculpture studio at Emerson College in Sussex, England, as a twenty-one-year-old biology student, I was naturally impressed by the seemingly haphazard collection of animal skulls and skeletons scattered around the room. Moreover, sketches of plants, minerals, seashells, and water vortices and meanders decorated the walls. In his quiet voice, Wilkes led the students into his main work studio, where barrels of clay sat among tangles of hoses, fiberglass molds, water pumps, and several-meter-wide clay basins. In this space, he spoke about water as a precious, life-sustaining element, permeated with vital qualities: a medium whose transparent innocence hides nothing, yet whose essential nature remains a mystery to us. Accordingly, water can be considered a substance worthy of veneration.

In our technologically developed world, however, we collectively tend to regard water in a utilitarian sense, as a commodity to satisfy the daily demands of our industrialized societies. To the degree that fresh water has become readily accessible, we seem to have lost an appreciation for this flowing medium upon which our very existence depends. As a student, I wondered whether I had ever paused to consider the nature of the liquid running so generously from the faucet. Its source? Its destination? Thus, through my studies with John Wilkes, the seed for my appreciation of water was planted, and unbeknownst to me at the time, I was taking my first steps toward an exploration of a dynamic way of thinking that was to weave through my later years.

In this essay, I examine water phenomenologically, that is, as it appears to us in its many guises. I return later to the work of John Wilkes after first drawing from his teacher and central inspiration: the hydrologist Theodor Schwenk (1910–1986), particularly his book *Sensitive Chaos*, a volume as singular as the liquid it portrays.[1] Schwenk served as director of the Institute for Flow Sciences in Herrischried, Germany, where he investigated qualitative aspects of fluid dynamics. Underlying all Schwenk's research was a recognition of water's close affinity to living processes, in that the existence of all life depends on water, and all organisms undergo a fluid phase during their development. Vertebrates, for example, pass through a fluid state during their early embryological development, and many unicellular aquatic organisms never fully "solidify" out of their watery environment and often bear resemblances to fluid forms, such as the vortex.

Schwenk maintained that a dedicated, comprehensive study of water would provide the observation and thinking necessary to guide researchers further in unraveling the secrets of the living world. "Through watching water... with unprejudiced eyes, our way of thinking becomes changed and more suited to the understanding of what is alive. This transformation of our way of thinking is... a decisive step that must be taken in the present day."[2] Schwenk's approach provides a path whereby one can enter holistically into a thoughtful, empathetic relationship with nature. By reacquainting ourselves with water in the way initiated by Schwenk, we can begin to revive an appreciation for this fluid upon which all life depends.

For a start, consider a geothermal spring amid a frozen winter landscape: a shell of ice draped upon adjacent rock formations; a clear, still, glassy-surfaced pool; and rising, metamorphosing clouds of steam. In such a place, water occurs simultaneously in all three of its earthly phases: solid, liquid, gas. Water manifests between the formed and the formless, the tangible and the intangible: the crystallized, hardened, geometrical qualities of ice and snowflakes, and the expansive, dissipating properties of vapor, respectively. By existing between the polarities of ice and vapor, liquid water exhibits characteristics of each, though these are expressed in dynamic qualities. Its formative capacity, for instance, is evident in the molding of landscapes and in supporting the growth and development of all living things. In contrast, water—the universal solvent—can dissolve solid matter effortlessly into formlessness and, through erosion, can level mountains. A river expresses both forming and dissolving tendencies through deposition and erosion, respectively.

Water illustrates an intrinsic polarity in another way. On the one hand, in the droplet, it shows a strong affinity to itself, a coalescence, a rounding off, self-enclosing gesture: a quality of formation. The sphere has long been a symbol of wholeness, of the universe, an independent, complete entity within itself. The Renaissance astronomer and mathematician Nicolaus Copernicus described the sphere as an integral whole that has no joints and is, in fact, the most perfect shape. A droplet of dew reflects a mirror image of the surrounding landscape of Earth and sky, makes visible the seven spectral colors, and is a tiny representation of the spherical, self-sustaining Earth.

On the other hand, water expresses a selflessness by "surrendering" its formative capacity and thereby adopting the shape of any vessel in which it is contained. Its ability to enclose itself within the sphere, and its proclivity to unite in dialogue with its surroundings, together bespeak the paradoxical nature of water. In this regard, I often enjoy looking into a shallow creek or pool, where I can either clearly see to the bottom, thanks to water's transparency, or with a slight shift of focus, the sediment and stones below vanish, and my gaze settles on a mirror reflection of surrounding trees and sky.

Water also displays two distinct qualities of movement: linear and circular. Gravitational forces influence water to move in a more or less straight line, as in falling raindrops, a cascading waterfall, and a steeply sloping stream cutting through a narrow valley. In contrast, an inclination toward circular formation is most patent in swirling vortices, expanding concentric rings created by raindrops impacting a still surface, bubbles, and the curls of breaking waves. The circular, rounding tendency of water is evident not only in the droplet, whirlpool, and associated phenomena but also in water's eroding activity. Flowing water, with the help of suspended abrasive materials, rounds off sharp edges and corners, as in stones at the seashore, and carves bowl-shaped depressions—called potholes—in the bedrock of a riverbed. In its affinity to roundness—the sphere—water offers a visual metaphor for wholeness and unity, as exemplified in the interrelationship of all earthly life, whose very existence is made possible by this aqueous medium.

Flowing water, as Schwenk indicates, "continually strives to return to its spherical form," but with the effect of the Earth's gravitational field, this tendency toward circularity can only partly be realized.[3] As mentioned earlier, the vortex is a circular phenomenon that occurs within a body of water where, in fact, the influence of gravity is mitigated. In a landscape, flowing water, in its inherent striving toward roundness counterposed with an externally imposed compulsion toward linearity, strikes a dynamic balance between these two tendencies in the meander. The meander, derived from the Greek *maiandros*, "a bend," shows a rhythmical swaying to and fro, which results in one of nature's most harmonious forms.

Schwenk observes that, in the meander, water's "endeavor to complete the circle is here only partially successful, as it cannot flow uphill back to its starting point. Right at the beginning of its circulatory movement, it is drawn downhill, and in following that downward pull, it swings alternately from side to side."[4] On almost level terrain, a river's meandering may become so accentuated as it strives toward circularity that some bends actually "pinch off" from the main stream, each forming an isolated oxbow lake.

Besides its expression in water, the meander can be seen in the vertically undulating flight of some birds, in the rhythmic slithering motions of snakes and salamanders, in swimming fishes, in the winding burrows of marine worms, and even in artifacts of human behavior. Concerning the latter, the architect P. J. Grillo observed that when a rectilinear cement walkway was obliterated by snow, pedestrians unwittingly forged a new path that had a graceful meandering quality; Grillo termed these naturally winding paths *slaloms*, after the similar curves traced on snow by downhill skiers.[5]

Another characteristic form given expression in water is the wave, and in wave formation too, water reveals inherent polarities. As described by Schwenk, in a wave at the seashore, beyond the breaker zone, it is the form that travels while the substance, that is, the water, remains stationary. Gulls floating on the ocean's surface bob up and down as swells pass beneath them; the birds' position is not displaced horizontally. (Imagine two people, each holding the end of a rope, with some slack, and one rapidly jerks their end; a wave will be generated, but the rope will not be displaced horizontally.)

In contrast, a standing wave created behind a stone in a shallow stream remains more or less stationary while water flows through it. A drifting leaf passes through the form, which remains fixed in space. Here, water reveals how form can arise solely out of movement, with undifferentiated substance streaming through. A parallel exists in living organisms: Substances are continually in transit, tissue cells arise and pass away, yet the organism's overall form remains stationary, that is, unchanged, save for gradual alterations due to aging.

Many cells in your body have been replaced during the past weeks, and others replaced several to ten years ago or longer—that is, cells have "flowed through" your form—but your overall appearance has probably changed little, considering the vast replacement of cells that has occurred. According to the biologist P. B. Medawar, "It is only the *form* of the body, the system of preferred stations for the inward-bound replacements, that achieves any kind of permanence at all."[6] A study of fluid dynamics can illuminate much about the genesis and maintenance of organic form, but the behavior of surface water on land is best understood through the action of that most dynamic of water phenomena: the river.

Interweaving currents, ephemeral vortices, swinging meanders, and turbulent churnings are just several of the many fluid qualities given expression in rivers. No other body of water exhibits such vital, indefatigable activities. In a very real sense, a river is the lifeblood of a landscape, and it not only affects local topography, geology, and biological communities but also shapes the cultural life of the people along its embankments. Their names alone stir vivid and diverse images of the landscapes and peoples, past and present, associated with them: Nile, Mississippi, Congo, Ganges, Yangtze, Thames, Rhine, Amazon. Despite their occurrence in widely distant lands, these great rivers, and rivers in general, share certain fundamental features.

The late Reinhard Koehler, a physicist who studied fluid dynamics in relation to biological processes at the C. G. Carus Institute in what is today Niefern-Öschelbronn, Germany, compared the unique qualities of each of the three anatomical parts of a typical river system: headwaters, midstream, and mouth. Here, too, opposite tendencies come to expression in water. For example, the precipitation-nourished headwaters are characterized by countless small, turbulent tributaries flowing at an accelerating pace over steep terrain, cascading over stones, vigorously eroding the land, and drawing together into a main stream in a gesture of confluence. There is little meandering in the headwaters; instead, a circular quality comes to expression in the creation of vortex trains downstream from innumerable stones.

In contrast, water at the mouth of a river, on relatively level terrain, decelerates, deposits its sediment load, and forms a delta as it enters and mixes with the sea. Here, the river branches into distributaries, demonstrating a gesture of spreading, expansion. Between these two extremes—contraction and expansion—appears the harmonious meander, which not only shows a striving toward balance, winding left then right through the landscape, but also exhibits an equilibrium between erosion and deposition, wearing away and up-building. Typically, while the deeper, outer bed of a meander is being eroded, sediment is being deposited on the shallower, inner curve, where the velocity is relatively slower; consequently, meanders are continually metamorphosing and translocating, reshaping the face of the land as they migrate across it.

In temperate latitudes, a seasonal rhythm is apparent in channel erosion and deposition. In spring and summer, the influx of meltwater washes over watersheds, picking up and transporting sediment. As the meltwater enters the river, water level rises, velocity increases, and the stream, with a full load of abrasive materials, maximally undercuts its banks. In autumn and winter, with reduced water input, the river's level falls, and it gradually decelerates.

Consequently, the stream's load comes to rest sequentially: stones and pebbles first, sand next, finer sediments last. One should bear in mind, however, that day-to-day variations, and even hourly changes, in a stream's activity, are superimposed over these general seasonal trends. In addition, a stream's behavior will alter along its course, reflecting variations in the nature of the bedrock, in the quality of the watershed's vegetation, and in general topography. Those who have the opportunity to observe even a small stretch of a stream through the year will be richly rewarded as one gets to know the "personality" of the flowing waters.

The down-cutting, eroding action of a river provides a striking example of what I call *simultaneous reciprocity*. As water carves a channel, and thereby sculpts the land, the channel, in turn, affects the stream by regulating its flow. Flowing water thus shapes its own vessel, which reciprocally influences the quality of fluid movement. The work of the stream on its channel, and of the channel on its stream, can be viewed as a dynamic whole functioning in space and time.

Here, again, there is a parallel to living organisms. During embryological development of the vertebrate circulatory system, blood begins to flow through the vessels when the heart has only just begun development, when it's literally not more than a simple tube. Can we consider that the heart, like a stream's channel, is formed by and bears the imprint of fluid movements and, in simultaneous reciprocity, regulates the flow of blood? According to Schwenk, "The fibers of the heart are a physical echo of the creative movements by which it was begotten. In spiraling paths, they swing down to its apex and then rise again to its base. They make the same movements and emphasize the revolving vortical streaming of the fluids within the heart."[7] Just as the developmental history of a stream and its channel can be seen as a dynamic unity, so can the blood and its "channel"—the heart and associated vessels.

We can extend the analogy further to view Earth, in its entirety, as an organism. In the hydrologic cycle—the "circulatory system" of Earth—water may circulate from sea to atmosphere to land and then back to sea, exemplifying an endless, unbroken circle. On its journey, a hypothetical drop of water may be incorporated into a magnificent cumulonimbus cloud, then become a six-pointed snowflake, then form part of a glacier, and next a clear mountain brook; subsequently, it may pass through plants, animals, and people. It is astonishing to imagine the potential circuit such a drop may take, but it is plausible that it could link together vast expanses of Earth and unify a multitude of life-forms, each being a temporary residence for our imaginary drop over eons of time.

But where lies the heart of Earth's circulatory system? Here, we can turn to the river for an answer. By examining its qualities, and by drawing on Reinhard Koehler's insights, it seems the heart of Earth's circulatory system is found in the river's meander. Just as the rhythmically beating heart resides in the middle region of a vertebrate's body—in the chest—the meander occupies the middle section of a river. And the river itself lies between its ultimate source in the clouds above and its final destination—the ocean—below at sea level, as expressed in Goethe's words: "A river from a cloud-wrapped chamber gone, / Of rock, and roaring to be one with ocean."[8]

A river's form, winding left and right through the expanses of a landscape, and its hourly, daily, and seasonal alternation of rising and falling water level, and of erosion and deposition, indicate a strong periodicity apparent in the rhythmic qualities of the heart. And as noted by Schwenk, every river generates its own individual rhythm: "The rhythm of its meanders is a part of the individual nature of a river. In a wide valley a river will swing in far-flung curves, whereas a narrow valley will cause it to wind to and fro in a 'faster' rhythm."[9]

For millennia, the rhythm of rivers played into the lives of people along the floodplains. There were times designated for planting crops in the silt deposited by receding floodwaters, times for irrigating fields, times for bathing, for collecting cooking water and for frolicking, times for fishing, and times for paying homage to the mighty powers of the flowing waters. Today, to a considerable degree, we have seemingly emancipated ourselves from nature's rhythms as epitomized by rivers. In fact, it is common for engineers to control and modify the cycle of the river itself. We build dams, erect levees, dredge riverbeds, divert enormous volumes of water for municipal use and massive irrigation projects, impound rivers, and even straighten channels to eliminate the inconvenience of lengthy, meandering shipping routes. Moreover, we all too often pump the refuse and chemical by-products of our civilization into these waters. We also neglect or deliberately deforest watersheds, resulting in inordinate volumes of surface runoff water and eroded soil entering streams, stressing them beyond their natural regulatory capabilities. As a consequence, many vessels of Earth's circulatory system are presently, if not imperiled, then significantly affected, which is exacerbated by the effects of climate change and resultant frequent regional flooding and droughts.

There are, however, efforts underway by governments and environmental organizations worldwide to counter some of the historical damages suffered by rivers and other freshwater bodies. The waters of the Thames and the Hudson Rivers, for example, are steadily returning to a condition in which they can once again support certain forms of aquatic life that vanished when these rivers became polluted.

Although all conservation and restoration efforts are necessary, in the long term it is crucial that our social attitude to water transform from viewing it as just a commodity to recognizing water as a *living* element. In this regard, I return briefly to the work studio of the sculptor John Wilkes, where I began this essay. After he described water as a life-sustaining element, he flipped on a switch that engaged a pump, and water began streaming through a series of clay vessels, cascading from one to another. As the students gathered closer, we could see how in each vessel water flowed in a figure-eight motion while rhythmically swaying from side to side. The movement was both hypnotic and soothing, and it was immediately apparent that, by developing these sculpted vessels, called "Flowforms," Wilkes has provided an aesthetic focus through which people can begin to appreciate water as a living element.[10] Accordingly, in the years since then, Wilkes and his colleagues around the world have done much to awaken an awareness of the vital qualities of water. As it swirls and pulses through the Flowforms, the streaming water reveals itself as a medium for rhythmical processes and, thus, as a medium to support life (fig. 1).

Figure 1. Flowform by John Wilkes, Vortex Garten Darmstadt; Wikimedia Commons, Creative Commons Attribution-Share Alike 3.0 Unported.

In closing, I recall Theodor Schwenk's wish: for us to overcome our preconceptions, how we habitually view the world, to arrive at a new perception of water. This is achieved when we struggle to awaken to a living appreciation of nature, to reforge a dynamic connection to the world around us. In doing so, the natural world takes on new meaning. Thus, a dewdrop on a blade of grass, a rain puddle on a woodland trail, a gurgling neighborhood creek, or a temporary vernal pond can become for us invitations to stimulate our imaginations, to enhance our "seeing," to begin an inward journey toward a new relationship to the natural world. And because water invites us to contemplate a world that exists between the tangible and the intangible, there is perhaps no better guide for our journey than water itself.

notes

1. Theodor Schwenk, *Sensitive Chaos: The Creation of Flowing Forms in Water and Air* (London: Rudolf Steiner Press, 1965).
2. Schwenk, *Sensitive Chaos*, 11.
3. Schwenk, *Sensitive Chaos*, 13.
4. Schwenk, *Sensitive Chaos*, 15.
5. Paul Jacques Grillo, *Form, Function, and Design* (Mineola, NY: Dover Publications, 1975).
6. P. B. Medawar, *The Uniqueness of the Individual* (New York: Basic Books, 1957), 109.
7. Schwenk, *Sensitive Chaos*, 91.
8. Johann Wolfgang von Goethe, "Immense Astonishment," in *Selected Poems*, ed. Christopher Middleton (Princeton, NJ: Princeton University Press, 1994), 177.
9. Schwenk, *Sensitive Chaos*, 15.
10. Mark Riegner and John Wilkes, "Flowforms and the Language of Water," in *Goethe's Way of Science: A Phenomenology of Nature*, ed. David Seamon and Arthur Zajonc (Albany, NY: SUNY Press, 1998), 233–52. See also John Wilkes, *Flowforms: The Rhythmic Power of Water* (Edinburgh, Scotland: Floris Books, 2003).

Been Ice

Robert Wrigley

July 10, 1976–October 16, 2022

Cold deposed. From underneath, lobes
and facets, a glacier belly silvergray
you rolled to your back to lick from time
to time, the crawl up wet and long:
ablation to terminus, up through downwasting
mud, grit, and stone; bergschrund, sun shaft,
plucked block, headwall—words you'd never heard before.
But from the foot of an unnamed icefield
a hundred yards long on a peak in Montana,
you peered beneath its lower lip, saw light from above,
and began with a seed's wonder
to rise up a slick of muck that last winter had been ice.
A climb that could have been done in two hours,
if you had not taken time to study caryatids of ice and stone,
weird columns that braced it all up: diminishing wrists
and elbowed arms, a swirling barber pole of silver and white
festooned with till and grit, the muzzle of a mule
with wolf front feet, a quartz and granite reef for a rump.
Four and a half hours of your first twenty-five years.
At one round and down-swelled dripping crystalline coupe,
ancient water dripped as cold as liquid light,
so cold it did not hurt and did not numb your teeth either,
but sizzled inside you like ice on a hot stove.
Today you have outlived it all.

Water Bearing Witness

Bruce Jennings

You don't look back along time but down through it, like water.
—Margaret Atwood, *Cat's Eye*

I grew up in a land chiseled flat by glaciers beneath an ice-age sky as they gouged the Great Lakes out. Today these deep reflective lakes are the residence of more than 20 percent of the planet's fresh water.[1] Consequently, two strong feelings are never far from me: the sublimity of water as a form of being and the moral obligation that water imposes on humanity to maintain a right relationship with it. Through direct experience and the mediation of cultural meaning and values, this is one way water has shaped me, and it is one of many ways that water shapes the humanity in us all.

What is most "elemental" in the elemental of water? For reasons that I hope will become clear, I am drawn to reflect on water's sensuality and movement. Our cognition of water is embodied, not merely representational and localized in the brain.[2] And no matter how we try to contain it—all the better to control and use it—water won't hold still. When I think of water in motion, the panorama I see includes the open ocean waves rising and falling like grasses in the wind on the prairie; the repetition of tidal ebb and flow, manifesting a force both eternal and extraterrestrial; and water's propensity to claim space (invading and colonizing, if you will) by flooding and overreaching banks and cliffs, boundaries imposed by the land through incarnational forces of its own. These trifles are merely temporary annoyances to water as it measures time.

Indeed, to interact with water is to be in motion, from the cellular to the planetary level. Recounting moments in my life when water has been an elemental force of healing, enabling, and remontancy is the final destination of this essay; bearing witness to water's protean power through memories is my route toward it.

Sensuality can be both inner and outer. In water's internal sensuality, I include the sensuousness of water on and in my body. Water is the medium and the method to move fluids among living cells. Water makes life itself as we know it on Earth possible. Evidence of ancient water on Mars suggests the possibility of ancient life. In its outer sensuality, I include the refreshment and purification of water's external touch. Outer sensuality is not so much a prerequisite of biological life as it is water's gift to living a certain quality of life. Because so many live in the world today deprived of regular and reliable access to fresh water—approximately two billion people, according to recent estimates—many of the potential benefits of water's outer sensuality have been denied to them.[3] Biology makes do, but justice is violated.

Water can be *thought* separately from the other elementals conceptually for certain purposes, but it cannot *be* separate ontologically from them. The water stories I have selected to tell have been thrust upon me by complex, intersecting elementals at work in my own life. Yet water deserves special attention. These stories have to do with what my mother gave me (touch), what my father gave me (solidity), and what my late wife gave me (companionship). They also have to do with what I have given them (fidelity). I am not yet finished with fidelity. It remains a work in progress.

What type of consciousness water may have, I cannot say. But it does bear witness. By that I do not mean merely being a voyeur or a bystander. By "witness" I mean an active, shaping presence quietly at work on humanity at all times, recognized or not. Water doesn't intend it; human beings are not aware of it. But there it is, nonetheless. In many ways, we are but sandstone hoodoos in the icy rain.

How Mothers Touch

Surely human healing has a stronger association with water than with any other elemental. Its cleansing and purifying properties are essential to the first-aid treatment of wounds that won't be fatal unless infection sets in. (The cauterizing heat of fire has also traditionally been used as a way of preventing blood loss and fatal infection.) More symbolically and psychologically, for centuries, certain locations or types of water—mineral springs, for example—have been associated with supernatural powers of healing. Pilgrims with chronic illness or disability make journeys to such sites. Some waters are more healing than others.

In the Christian tradition, the ritual of baptism is a theologically complex example of the role of water in renewal, rebirth, and forgiveness. In the Gospel story of one of the miracles performed by Jesus, he encounters a sick man sitting beside the waters of a well in Jerusalem well known for its curative powers. "When Jesus saw him and knew that he had been lying there a long time, he said to him, 'Do you want to be healed?' The sick man answered him, 'Sir, I have no man to put me into the pool....' Jesus said to him, 'Rise, take up your pallet, and walk'" (John 5: 6–8).

When I was one year old in 1950, I contracted poliomyelitis. At the time there was a serious epidemic of this devastating, highly contagious disease going on, and an effective vaccine to protect against polio would not be developed and brought into use until about 1955.[4] Too late for me.

My mother and father had gone on a trip to the Canadian Rockies, a kind of second honeymoon, and left me home in Indiana in the care of my aunt. When they returned, she reported that I had gotten a flulike illness while they were away, but it had passed. At that age, I was just beginning to take steps and walk—one of many amazing developmental periods along the pathway of human becoming. I had been toddling well before their trip and my illness, but I began to fall down. My parents called in the family doctor. He came to our house, and he gave them the grim news. As my diagnosis matured, the polio seemed to be affecting mainly my left leg. The probable location of the nerve damage was such that muscle impairment would not spread to diaphragm and chest, so the prospect of a lifetime confinement in an iron lung was off the table. But my infection could have brought about a permanent functional impairment of one of my legs.

The treatment for me at that time was hydrotherapy to stimulate damaged nerves and muscles in my leg so that they might regenerate and I would be able to walk normally. Thus it was that my life was taken over by daily sessions in a basement room of the local hospital equipped with large whirlpool tanks, whose strong water jets would buffet my leg in an effort to "wake it up," as my mother and I later came to think about it. A nurse told my mother that the more running-water stimulation I could receive, the better my chances of walking normally. So my mother would bring me home and continue my therapy at the kitchen sink with the faucet running onto my leg in the basin beneath it. During those times, my mother would sing quietly to my leg as she kneaded it.

The hospital sessions lasted for hours, and there were so many afflicted children at that time that the hospital facility was fully used. I was too young to socialize, but my mother struck up lifelong relationships with many of the mothers who were there. In later years, they would get together and my mother would see which children had become impaired and which had not.

Most of my recollections of that time come from what my mother and father told me much later, of course. But still, I think I do have some direct memories, fragmented and fuzzy as they are. The hydrotherapy room and the whirlpool baths, where as a toddler in a diaper I was baptized several times a day for many, many days, is an original image memory for me, not a later reconstruction. I hold on to that memory viscerally, as something I know to be an actual part of my young life. Something that I need to be true.

What I remember first and foremost is touch, or better, touch that carried presence with it. I have tried to write the closing section of this story in a deliberately disjointed way. I do this to capture what the water saw or, what amounts to the same thing, what I see in memory reflected in the water—the very water that was keeping my mother alive as a living organism and that was pulsing rhythmically against my leg to save it from the trespass of the virus.

The feel of the warm water against the skin of my leg. Returning touch on my bad leg, two hands, moving, grasping. Gentle pressure kneading and needing, muscles feebly feeling feeling, that was the daily prayer.

I was so young, just walking, just go-getting when one of my legs said no. The command brought her there lending her hands laying them on that leg singing softly to it each day.

I was so young. Could I remember this? In memory or dream I was there. My mother was there. Touching. Touching with the water. That was enough. Dangling leg hands laid on, as many times as it took.

"Get up and walk."

After the hiatus, some did not; I did.

I was so young. Later life painted pictures of lingering touching of my leg and my mind. But ignore the pictures. The touch of my mother and the water were the reality that mattered. For she and the water were always there.

Casting upon Waters

When my father died of pneumonia in 1989, the specific timing of his death was determined by a deliberate decision to forgo life-sustaining medical treatment. In my father's case, our family—my mother and I, for I am an only child—had seen this coming for a long time.

Profoundly demented with advanced Alzheimer's, incontinent, noncommunicative, often combative, Dad had resided in a nursing home for three years. He had needed my mother's gradually intensifying care and supervision at home for about eight years before that. Giving that care nearly did her in; it was a brutal closing decade to an otherwise generally happy and prosperous forty-year life together in marriage. Her dance with my polio was short by comparison.

Thinking back over my childhood and my interactions with my father as a child, what stand out as the best times were spent at a weekend cottage we owned on the shore of Lake Wawasee in northeastern Indiana. My father was an outdoorsman and enjoyed hunting and fishing, especially still-water fishing for bass in the lake. My memories are vivid of going out in our boat, learning the lake and its underwater secrets, particularly holes where the largemouth bass liked to feed. My father would cast food—and death—for them out onto the waters of the lake. Trolling was another way to fish, and the early morning and late afternoon seemed to particularly appeal to my father. With rather heavy eyelids many times, I would accompany him. Now I am grateful to him for those times, especially for the quiet, even the silence, of the lake at that hour. The red-winged blackbirds, my favorite callers, were responsive to it. Who are you? Where are you? On the lake and elsewhere, my father taught me how to answer those questions; that is, he gave me confidence, competence, and belonging—the whos and wheres of my life.

A lifetime later, I would visit him in his nursing home. Sometimes he recognized me, sometimes not. But it was clear when he was communing with a long-term memory that would absorb him. The temporal balance of his life had shifted, and I inferred from his demeanor that he was living from those memories of the past more than he was from the immediate experiences of his present.

Accordingly, as I sat with him, I often constructed a rather elaborate fantasy. First, about the kind of togetherness we had on the lake thirty years before. Second, about the kind of togetherness we might have if both of us happened to get caught up in the same memory at the same time. Memory is liquid and it pooled that day when what he was remembering in a given moment happened to overlap with my recollection of that same moment. Our shared past became present, as a body of water is replenished by new water flowing in from the surrounding watershed. We need a name for this. Perhaps *memoryshed* would do.

Here is a story of my visit with my father in his nursing home one afternoon.

My father once, and still, serenaded by cicadas at sunset, in the dayroom on a plastic chair he sits, despumating intentions. Buttons closed at the cuff encircle hands once thick as slabs. Emissaries of a man, those hands, that once roamed free to caress a woman's cheek. He glances at them, then looks away, leaving them for me to study.

At length, he decides to stand. Fingers now slim slide forward on chrome arms; he presses legs and hips down hard against the chair, measuring with a crooked sextant the angle of his repose and the constellations, the figures, of his age.

He eases back down and watches himself as a young boy skating on the frozen pond behind the barn at the Ontario place, blue-glazed linoleum at dawn. "Hughie," his mother calls to the boy from the house, "come in for breakfast, Hugh."

Forgetting that he forgot, he returns to an unfinished task. Hands climbing arms, he strains to rise. Just then, the sun falling on him, he sinks back squinting: another memory passing by.

The water of Lake Wawasee is clear in the shallows. On the dock he shows a boy of six how to bait a hook. Feel the peat. Separate worm from worm. Looking down, they can see sunnies and bluegills darting. Bobbing myself on eddies of memory that spill into the same pool, I am with him again. I am that boy, and he is that man. I smell moss and separate worm from worm, untangling them to squirm useful on barbed hooks. Standing together there, we were reflected in the lake water, and I almost think we might have taken note of our reflections at that moment.

Aldo Leopold asked what a mountain would think of what he had done when he shot a wolf. Let us ask this question of water, as well as of the land. If cameras record moments and memories record moments, then I suspect that lakes record them too. It is merely that what the lake sees and remembers has to be accessed in a different way. As a memory of a memory? Or as a physical visit to the lake and our dock (no doubt long gone) later in life? If only we knew how to develop and view what the waters of the earth have seen.

Quiet Waters

I turn now to two encounters with water in the long life together my wife and I built. Swimming and making a home.

The first story, "Beautiful Swimmers," has to do with our shared passion for swimming and the safety and danger that water always carries with it as an elemental force in our lives. In the case of the first story at least, I must give due credit to the air, for it plays as important a role as water in this story. Air in the guise of a sudden thunderstorm and deep water converged in a way that happened to offer my wife and me an important insight about each other.

Similarly, the second story, "Silent Snow," considers the forms or states of water. Water in solid form can be as creatively constructing and communicative as it is in its liquid state. As for gaseous water, think of how the harnessing of the power of steam by the invention of an efficient steam engine in the eighteenth century changed the world.

Beautiful Swimmers

She and I swam a long way in our crossing of the big lake. Middle of the journey—shore behind barely discernible, a thin line in the haze; shore ahead, little more. We were beautiful swimmers. Of the solitary things we did together, it was perhaps the best.

We traveled light, smooth muscles worked slowly, no wasted motion, efficiently gliding through the lake water. Water rippling, our breath coming in and going out. We carried only wonder and fear and wore little cloth covers, useless for the swimming, and needed only for the propriety of returning to land. We both would have dared to go without them.

When we return to land. *If* we return to land. The weather was coming, many boats were around, one would get us. The time was for waiting. I floated on my back, watching the dark clouds form. She clung hard to a buoy and stared down its slimy chain into dark water.

Silent Snow

I don't know why the snow bothers to fall at all, it is so widely unappreciated. When it lays down five or six inches overnight, few awaken to a sense of its beauty and peace. Instead, most are irate at the inconvenience. They all have other places to go in order to be somewhere.

Living in a house with a long driveway to the public street and in a frosty winter climate with frequent deep snowfalls, my wife and I devised the following routine. She and I would arise before dawn to shovel a path down the driveway so we could drive away later, after the plows had come and cleared the street. We did not wait for the snow to stop falling, for then it might be deep and heavy and a chore to move. Better to stay ahead of it with repeated returns to the shovels and the task.

At that hour, we had the world all to ourselves in a cold that could be touched as well as felt. We were in a togetherness that was a gift as silent and as secret as the snow itself. Our loneliness overcome, the snow partook of that with us, and each descending flake seemed less alone.

Most people say words like *drifting, heavy, light, dry powder* for skiing, or *wet packing* for snowmen or snowball fights. After shoveling, we reflected on something else over cups of hot cocoa. The essence of snow, we agreed, is the silence of its falling and its secrets whispered to us as we threw shovelfuls of it into the clearing of the world.

Thus, we used to work gloved and muffled in the darkness of cold air and watched the light in the east come up over the trees and the roof of the house that was our home.

Laying to Rest Promises Kept

When she died in 2020, my wife, Maggie Jennings, was cremated in accordance with her wishes and instructions. Under different circumstances, we might have talked about more green alternatives to conventional cremation, but she was getting too weak, and I was already too exhausted to navigate whatever would have been required to avail ourselves of those alternatives.[5] So now her ashes sit on a shelf in the back of my closet. They await a discrete spreading at a gathering of friends and family in a spot she loved, overlooking the Hudson River and the Palisades cliffs on the other side. The morning sun striking those cliffs triggers an inner glow emanating from the rocks in a most uncanny way. Glows from within were what my wife liked best, in geology and in people.

She was an Oklahoma girl who went to college in the Hudson Valley and then lived almost all her adult life there. I can't speak for other rivers, but the Hudson River does talk to you almost daily in the slapping on the shoreline of its waves stirred up by the tides or the passing of an enormous freighter or barge hauling their wares to the insatiable mouth of New York City.[6] It persuades you to allow yourself to be lost in a place where the lost are remembered. She was clear, convinced, and not afraid. I was the one ambivalent and uncertain.

My family and cultural tradition, my people, self-scrutinize to excess. My thoughts, even after three years, keep asking me how well I took care of her. True, I stayed, I persevered, I did not jump ship like some husbands do when their world comes crashing down on their heads. When we were waiting to be rescued in the middle of the lake when a storm came up, I would never have swum off and left her holding on to the buoy. Never. That's not the issue. But was my presence during her illness a healing thing the way I offered it? Being taken care of, even by someone you love, especially by someone you love, is a heavy thing to bear.

In all our time together, truly missing from the marrow of my life was the courage to give myself to others in a situation without remainder, without holding something back. In our life together I withheld something, even from her. I came to believe that, in her final illness, I would find at last the courage to commit everything. "Sell everything that you have, and follow me," as it is said.

On the day she died, my family gathered in her room at the hospice residential facility, and she lay on the bed without the motion of water or anything else. At a certain point, a hospice nurse stepped everyone out of the room before the men from the funeral home went in with their gurney. Before long, they rolled her into the hallway and turned left toward the back, where the hearse was waiting, while my family turned right toward the front. I had to choose.

Without knowing how, I chose—I looked left and turned right to follow my son and turned my back on her.

I think of this as a crisis of fidelity. I sit in judgment on that moment now and ask whether I should have gone with her to the crematorium, where she would be so alone. I say to myself that it is better to accompany those loved toward the future than to follow one loved toward the past. But I am not sure. I think I never will be.

There is a second aspect of this crisis of fidelity. It has to do with a pending ceremony, which I cannot put off forever. Despite my wife's wishes, I am troubled by the prospect of putting her remains into the flow of the mighty river where they will dissipate and be swept away by the ever-moving waters. It is the placelessness of her final emplacement that troubles me.

True, in future years I can stand at that spot by the river and know where she is—in the abstract, out-there, circle-of-life sense. Indeed, I think it is in the nature of the elemental water to negate a more concrete locale of emplacement. In contrast, the land, Terra, does provide for that tangibility and offers a different kind of solace and memory. I find myself wishing to get a proper urn and a plot in a cemetery, where her remains could lie down among her family or mine. And where I could go in my remaining years and look at a small, engraved metal plaque on the ground identifying her. In other words, a concrete place I could visit with my body as well as my mind, a body kneeling down to touch the soil and the grass in ways that touching the river is not possible.

And Promises Yet to Keep

It was fidelity that bonded Maggie and me with one another while she was alive. It is fidelity also that now bonds me to her, to my mother, and to my father via memory. Water exercises fidelity too; all those properties of being that tradition has regarded as elemental do so. Perhaps water's greatest fidelity is the gift of possibility; water makes us possible in myriad ways. Included among the human possibilities for which water provides the foundation are our love, care, concern, respect, and trust for one another.

My fidelity to Maggie, I now realize, is fidelity to water. And the respect and awe I feel for water are themselves a reflection of the respect and awe I had—and have—for her. If I return Maggie's ashes to the waters, I may not have a memorial plot, but I will have a place to hold memories even as they—my memories and her ashes—are always in motion. Like the water, her spirit is *esemplastic*, to borrow a term from the philosopher-poet Samuel Taylor Coleridge. I think of her now as swimming laps, ever touching and feeling the water's touch, ever making things whole, ever sustaining the world.

As for me, I shall keep on looking for the right current to join as I move in her direction.

notes

1. "Keeping the Great Lakes' Fresh Water Clean Is a Tall Order," Office of Response and Restoration, National Oceanic and Atmospheric Administration, 2022, https://response.restoration.noaa.gov/about/media/keeping-great-lakes-freshwater-clean-tall-order.html.
2. A. Noë, *Out of Our Heads: Why You Are Not Your Brain, and Other Lessons from the Biology of Consciousness* (New York: Hill and Wang, 2010).
3. UNESCO World Water Assessment Programme, *United Nations World Water Development Report: 2023: Partnerships and Cooperation for Water* (Paris: UNESCO), https://unesdoc.unesco.org/ark:/48223/pf0000384655.
4. On the history of polio, see the Mayo Clinic, "1955: Polio," https://www.mayoclinic.org/coronavirus-covid-19/history-disease-outbreaks-vaccine-timeline/polio.
5. As it happened, I had contributed a chapter in a book on this very subject: B. Jennings and C. M. Hammack-Aviran, "A Covenant among Generations: Keeping Trust in the Republic and the Law," in *The Future of the Corpse: Our Changing Ecologies of Death and Disposition*, ed. C. Stoudt and K. Rothstein (New York: ABC-CLIO, 2021), 198–225. Writing an essay on what is ethical is one thing; finding the strength to do what you know would be ethically right when you are trying to cope with impending death is another.
6. Technically, the Hudson is a tidal estuary and a fjord rather than a river. Its tidal action extends as far north as Troy, New York.

Lake Return

CD Wright

Maybe you have to be from there to hear it sing:
Give me your waterweeds, your nipples,
your shoehorn and your four-year letter jacket,
the molded leftovers from the singed pot.
Now let me see your underside, white as fishes.
I lower my gaze against your clitoral light.

The Big O

Hannah Close

*T*he swell intensifies. Beads of salt water gather and glisten on the surface of my skin as the upwelling approaches. My body shivers with each surge, its edges under pressure from the penetrating forces that pour over it. Naked, I am pushed to and fro in a restless exchange, my boundary indiscernible from the other. Skin to skin, shore to shore, we spin together through the cosmos in liquid rapture. A wave of pleasure overcomes me. For an eternal moment, I sink beneath the veil, submerged in euphoria. Eventually, I float to the surface, catching my breath as I rise. My body—soft, porous—oozes with the world around it, dissolved, together with the other, in a sublime watery continuum.

The ocean is eros writ large. Eros, or the erotic, is an elemental force for life that operates within and beyond the sexual. Eros is elemental because, like the mineral and classical elements historically regarded as fundamental to life processes, it precedes and pervades all that is. It is an essential building block of life. The ocean comprises one of the classical elements, water, and is the origin of Earth's first life-form. Its influence pervades even the most remote mountain valleys, as observed in the drifting diatoms wafted along by roiling ocean winds, eventually settling on alpine slopes to nourish the local flora. Without the ocean, and without the unifying, integrative force of eros, there would be no life. Like eros, the ocean connects us all. It is the aqueous, organic glue that brings us together across borders both real and imagined. Like an arrow from Cupid's bow, eros and the ocean traverse distances in order to unite life with itself. As the Canadian poet and scholar of eros Anne Carson says: "Desire moves. Eros is a verb."[1]

In Greek mythology, Eros is the god of love and desire, responsible for procreation and passionate entanglements of all kinds. Eros inspires novel possibilities for vitality and aliveness through the unity of lovers, human and otherwise. In many ancient myths and religions, the ocean is seen as the source of all life, the place from which everything else emerges. It is a domain of boundless creativity and fecundity. Through the frame of Greek myth, Eros is very much at work in this liquescent realm. Like the womb of the world, the ocean is forever giving birth to new creatures, islands, and stories, her belly full of nascent life forms. From trillions of microscopic algae to pods of majestic blue whales, the largest mammals on the planet, the ocean is one of our most fertile and productive ecosystems.

The enlivening power of eros comes to the fore through touch, the "contact between bodies," and the yoking together and expansion of aliveness between beings.[2] Eros is sensual, pertaining to the senses and to the embodied, tactile, and felt aspects of reality. This includes the feltness and physicality of thoughts and emotions, to which many falsely ascribe the label "abstract" or "disembodied," hence "unecological." As a kind of organic intelligence rooted in reciprocity, eros recognizes that awareness, physical or otherwise, originates from the earth beneath our feet, both watery and solid, which in turn originates from the cosmic expanse that envelops our planet. In this vein, what could be more "natural" than having our "head in the clouds"? The clouds, after all, are made from the oceans through which we glide, perched upon rickety wooden vessels that glug lazily amid the swell, guided by the stars above.

In this sense, eros bypasses reductive categories and beckons toward a more integrative view of the world that ultimately nourishes life. Eros is, in essence, a relational force. It is a force of attraction, a kind of primordial magnetism. This is what I mean when I say it is integrative. It recognizes kinship between distinct life-forms, which includes beings, ideas, and feelings, and draws them together. In drawing together distinct life-forms, however, eros requires that they are indeed that: distinct, separate. Eros adheres to the paradoxical maxim put forward by the French philosopher and mystic Simone Weil that "every separation is also a link."[3] To knit life together, eros must have access to unique threads, lifelines that spool out individually on their way to entanglement.

Despite this fundamental need for separateness, *eros* is ultimately an agent of proximity. It is in the tantalizingly close that we feel eros at its most powerful, where those in relation become so close as to almost extinguish the other, but without dampening the flame of aliveness expressed in each individual. Intimate distances form erotic terrain, with lines of relationship intersecting to create an animate topography of taction, immanence, and betweenness. In this sense, I use the term eros to describe the mediating, catalyzing force that expands and contracts the potency of relation between beings. Like an electrical current, eros oscillates along a spectrum of polarity, amplifying difference and the unique expressivity and diversity of living beings while integrating them into a container of oneness in the same shared breath. The ocean is, of course, separate from us, but it is also not. Salt water runs through the tributaries of our circulatory system as much as it fills out the great terraqueous basin that encircles our planet. In our embodied "islandness," we too emerge from the ocean's abyssal depths as forms amid apparent formlessness. While the ocean might be distinct from us and from the land that punctuates its seemingly infinite flow, the ocean as "eros writ large" provokes all of these questions and more around proximity, porosity, and entanglement.

I'd like to include a brief note on metaphor before diving into this topic further. Like the ocean and eros, metaphor precedes and pervades not only this particular essay but also much of life itself. While stimulated in part by a nonconceptual environment, our consciousness is embellished with metaphorical understanding.

It may come as no surprise, then, that many of the metaphors we use to describe eros and erotic sensations, as well as other physical or emotional sensations, are inspired by the ocean. Metaphor can be a powerful means for connecting with the ocean's erotic power and, from that, our own erotic power. It allows us to carry insights into the nature of reality from the material into the metaphysical, and vice versa, recognizing that the line between these domains is porous and that neither one is more true or more ecological than the other.

Metaphor is, contrary to attributions of "pure rhetoric" or "mere abstraction," a living, embodied, relational phenomenon. We might even say that it grows out of the ocean floor like a succulent stem of sargassum seaweed. It is itself erotic. It is so because it is about forming relationships and recognizing kinship between domains, much like the ocean forms connections between islands, currents, seas, and pelagic layers. In literary or psychological terms, one domain is mapped onto another to draw out their semantic similarities, their conceptual kinship, if you will; kinship grounded in meaning-full connections that bridge the gap between ignorance and understanding. Metaphor bypasses a lack of rational likenesses to extend our imaginations beyond the limitations of standalone concepts and, like the ocean, toward a continuum of connection.

Metaphor means "the bearer of the beyond" or "to carry across." Like eros and the ocean, metaphor takes us "beyond" into new realms of sensation, imagination, and aliveness. Metaphor, metaphorically speaking, is also the vessel that carries us across uncharted waters as we navigate our individual and shared subconscious, realms frequently associated with eros and the ocean. It is with this awareness of metaphor that I encourage you to engage with the following.

You may have noticed that the opening paragraph depicted both an orgasm and an encounter with the ocean. As mentioned, eros contains the sexual but goes beyond it. *Intercourse* need not refer only to the meeting of genitals. Still, sex is an essential part of life; indeed, it is a primary source of it, much like the aforementioned elements that constitute our world. The French poet Paul Valéry made an explicit connection between sex and the ocean when he described swimming as "fornication avec l'onde," or fornication with the waves.[4] The ocean applies pressure in varying degrees to the body, as a lover might do during sexual intercourse. It inevitably touches every inch of us as we encounter it. It is almost promiscuous in its caress, insatiable and insistent on absolute intimacy and exposure. Its viscosity becomes contiguous with ours. Salty liquids osmose between our body of water and its body of water. Its surface courses over the skin of our body as our surface, correspondingly, courses over the skin of the ocean. Like the "big O," we rise and fall inside this field of acute physical sensation, following the crest and crescendo of intimate bodily feeling. Incidentally, *scend* in cre*scend*o is an archaic term for the rising motion of a wave, and the broader meaning of *crescendo* is "to grow," which gestures toward the feeling of expansiveness we feel with both the ocean and our lover when we oscillate together in this embodied exchange.

This feeling of actual and symbolic sexual union can lead to an enlargement of the self, including, sometimes, the enlargement of your genetic imprint in the world via children but also an enlargement of the self through an awareness of the self in the other, kin to kin, salt-stippled atoms to salt-stippled atoms. As the Irish poet John O'Donohue noted, "At the ecstasy point in making love, the language is again the liquid language or release of self into other."[5] The French philosopher George Bataille also recognized such an event as metaphorically relative to the ocean: "Erotic activity, by dissolving the separate beings that participate in it, reveals their fundamental continuity, like the waves of a stormy sea."[6] Eventually, this exchange may compel you to recognize yourself in your lover and in the waters that surround you. Your body is approximately 60 percent water, nearly the same proportion of ocean that covers Earth's surface, highlighting an obvious physical affinity. What's more, the ocean, like a vast pelagic mirror, reflects back to you the unfathomable depths of your subconscious, inviting you to gaze ever more deeply into the self as you encounter the other. Self-absorption without ego, perhaps. This subconscious domain is not really yours but is instead part of a commons of immanent awareness shared with the animate world around and within you. Intense romantic union often reveals this fundamental kinship to us in ways that are unsettling to the individualistic ego due to the immense vulnerability required to initiate such an intimate relationship. In the same breath, the ultimate knowledge that we are not, in the end, alone, is the promise that keeps us alive. The ocean serves to remind us of this absolute, life-affirming connectedness.

Despite eros being identified with a male god in Greek myth, erotic power is more commonly associated with women and their capacity for birth, as well as their connection with seduction and beauty. However, it may be more apt to associate eros with the feminine in some cases. In some versions of the Greek myth, Eros is the son of Aphrodite, the goddess of sexual love and beauty, born from the "foam of the sea" created by the severed genitals of Uranus after his son Cronus threw them into the ocean. In the painting *The Birth of Venus* (the Roman version of Aphrodite) by Sandro Botticelli, the goddess emerges naked from the ocean on a shell. This image illustrates the idea that eros is a natural force that emerges from the depths of the unconscious and ocean alike, and ultimately from, or rather through, a woman. However, it would be incorrect to say that eros is not active in men. Without their equally important life-giving capacities—metaphorically contiguous with the ejaculatory trajectory of Eros's arrow and the salt water that constitutes the ocean—birth, hence life, would not be possible.

Other female ocean and water deities such as the Celtic Clíodhna, the Yoruba Oshun, and the Greek Calypso, Charybdis, and Scylla also depict erotically charged figures (note, they are not oversexualized or sterilized in a pornographic sense) who wield the mighty power of the ocean. Despite the association with life and birth, ocean goddesses are frequently depicted as sirens, temptresses, and originators of violent tempests, an encounter with which often proves fatal to male sailors. As Anne Carson writes, "An influx like eros becomes a concrete personal threat."[7] In this vein, both the ocean and the feminine, as agents of the erotic and symbolic of the vast unknown, represent something that is paradoxically feared and desired by the male psyche, which sets out to conquer and contain this domain of unfathomable elemental wilderness. Predictably, this fails. Like the ocean slipping through the palm of your hand, this attempt at capture scatters eros to the wind: "Eros moves. You reach. Eros is gone."[8] This subterranean "something" that eludes capture represents a form of consciousness or aliveness that is not meant to be directly perceived, despite being active in the world. Once again, this is how metaphor itself has affinities with eros; it is a form of indirect awareness that yields its magic only in the instance that it is not grasped too forcefully but is instead permitted to flourish in the margins of our perception, embellishing it with shades of life and meaning.

The oftentimes seductive draw to death, as symbolized by the siren archetype, is also, paradoxically, an invitation toward life. Indeed, the orgasm, the point at which life is often conceived in literal terms, is also known as *la petite mort*, or "the little death." It marks the threshold point between this world and the next at which the tenuous boundary of contained consciousness is compromised. As George Bataille notes, "The urge towards love, pushed to its limit, is an urge toward death."[9] The inverse is also true; the urge toward death is, in some cases, an urge toward love, eros, and life. Should your ship founder and you drown, your body eventually falls to the seabed, where it decomposes and becomes food for other creatures, who absorb your self into their selves so that their lives can continue. In this vein, you prevail in the world by nourishing others. Death leads to life, which leads to death, and so this eternal cycle of bodies emerging and dissolving amid the frothing ocean waves continues in the spirit of eros. For Andreas Weber, the biophilosopher and originator of the term *erotic ecology*: "If we want to love, we must learn to carry forth this Eros. We must give a part of our own ecstatic individuation back to the world, must therefore be prepared to die, to some extent, so that something else might be."[10]

Tales of sinking ships, whirlpools, and demonic monsters of the deep associated with these goddesses also emphasize the unpredictability and loss of control associated with the ocean and eros. These metaphors are often linked with stereotypes that depict women as "out of control" and "overemotional," particularly around the time of menstruation, which, like the ocean tides, has a strong symbolic connection with the moon. These surges of feeling represent a dangerous kind of volatility to hyperrational, patriarchal societies that rely on a sense of fixity and submissiveness to enforce dominance. Similarly, "waters breaking" during birth is indicative of this loss of control. After months of containment and concealment, new life bursts forth somewhat violently, bringing with it salt water, flesh, blood, and the confronting wailing of new consciousness. Like the ocean disgorging itself of creatures seasonally destined for the dunes, it also brings what is inside *out*, presenting the world with a form of consciousness that has, up until this point, remained hidden under the surface of the mother's skin.

So, although the ocean and eros are reminders of our own mortality and the fleeting nature of human existence, they are also symbols of hope and possibility, offering us a glimpse of something greater than ourselves, something that is both beautiful and terrifying in the same breath. The vastness and mystery of the ocean can inspire feelings of awe and wonder but also a sense of insignificance and a longing for something beyond ourselves. Similarly, eros can inspire a powerful desire for connection and intimacy but also a sense of longing and frustration when those desires are unfulfilled. In both cases, there is a sense of yearning for something that is just out of reach, a feeling that is both exhilarating and painful. This sense of being on the edge of something, an almost excruciating immanence, is, paradoxically, one of the most enlivening things we can experience. Such an experience draws us forward into aliveness and into the aliveness of others. It is a leap toward relation and, in turn, toward life. Like the bow of a boat slowly descending into the trough of a colossal wave, with gravity as its anchor and time almost standing still, the moment in which we come close to being overturned, or to merging with what's around us, depending on how you look at it, is the moment in which we are most in touch with the elemental eros of being.

notes

1. Anne Carson, *Eros the Bittersweet* (Dallas, TX: Dalkey Archive Press, 2022), 19.
2. Andreas Weber, *Matter and Desire: An Erotic Ecology* (White River Junction, VT: Chelsea Green Publishing, 2017), 23.
3. Simone Weil, *Gravity and Grace* (Lincoln: University of Nebraska Press, 1952), 200.
4. Charles Sprawson, *Haunts of the Black Masseur: The Swimmer as Hero* (New York: Pantheon Books, 1992).
5. John O'Donohue, *The Four Elements* (Dublin: Transworld Ireland, 2012), 53.
6. George Bataille, *Eroticism* (New York: Penguin Classics, 2012), 22.
7. Carson, *Eros*, 49.
8. Carson, *Eros*, 185.
9. Bataille, *Eroticism*, 42.
10. Weber, *Matter and Desire*, 60.

Sea: Night Surfing in Bolinas

Forrest Gander

Maybe enough light • to score a wave • reflecting
moonlight, sand • reflecting moonlight and
you • spotting from shore • what you see only •
as silhouette against detonating bands • of blue-white
effervescence • when the crown of the falling • swell
explodes upward • as the underwave blows through
it • a flash of visibility quickly • snuffed by
night • the surf fizzling and churning • remitting itself
to darkness • with a violent stertor • in competition
with no other sounds

paddling through dicey backwash • the break
zone of • waist-high NW swell • as into a wall of
obsidian • indistinguishable from night sky • diving
under, paddling fast • and before I sit • one arm
over my board • I duck and • listen a moment
underwater • to the alien soundscape • not daytime's
clicks and whines of • ship engines and sonar • but
a low-spectrum hum • the acoustic signature of fish,
squid, • crustaceans rising en masse • to feed at
the surface I feel • an eerie • peacefulness veined
with fear

after twenty minutes the eyes • adjust, behind
the hand dragging through water • bioluminescent
trails • not enough light • to spot boils • or
flaws in nearing • waves appear even larger • closing-in
fast • then five short strokes into a dimensionless •
peeler, two S-shaped turns, the • kick out, and from
shore • your shout

it is cowardice that turns my eyes • from the now-empty
beach • for with you I became • aware of an exceptional
chance • I don't believe in • objective description,
only • this mess, experience, the perceived • world
sometimes shared • in which life doesn't • endure,
only • the void endures • but your vitality stirred
it • leaving trails of excitation • I've risen from the
bottom of • myself to find • I exist in you • exist in
me and • against odds I've known even rapture, • rare
event, • which calls for • but one witness

Walking with the Invisible

Marzieh Miri

Sit down beside the stream sometime and watch
Life flow past. That brief hint of this world
That passes by so swiftly for us is enough.
    —Hafez, "Divan of Poems," translated by Robert Bly and
    Leonard Lewisohn

Back home in Iran, water is scarce. The ancient city of Shiraz is surrounded by mountains and has a seasonal river named the Dry River. The river flows only in winter or spring. Water comes from the mountains and creates tiny creeks around the city, thereby fashioning pleasant picnic spots. The scarcity of water causes people to cherish these temporary streams—to sit by them and watch the flow of their temporary lives.

From the perspective of someone who grew up in Iran, it is odd to me that anyone would bury a river—how ungrateful. That is why, I think, I became interested in looking for Taddle Creek—a lost river in my new home, Toronto, Canada.

Taddle Creek is a buried stream that flows about six kilometers through the downtown core. Eventually, it empties into Lake Ontario—one of the five Great Lakes that straddles the Canadian-US border. Originally a gathering place for local Indigenous communities, the stream served as both a life-giving and a spiritual source for the Huron-Wendat peoples. But by 1884, as the city grew and the waters became a receptacle for waste disposal, like many urban rivers, Taddle Creek was buried as part of the city's underground sewer system.

A few years ago, this waterway became the subject of my graduate thesis that explored the course of the creek, specifically as a lost landscape. My project was a search for a buried piece of nature and Indigenous history and an attempt to retrieve the meaning of these lost moments through a multimedia installation. As an architect and photographer, I have been researching and creating work on the theme of sense of place and the relationship between humans and the environment for years. This project became part of a larger journey to understand how we compromise nature in the city and how we may eventually rethink the costs of so-called progress.

Having immigrated to Canada in late 2019, the timing of this work coincided with the emergence of the COVID-19 pandemic. Fortunately, the Taddle Creek project became a source of health and healing for me. Certainly, for many people, coping with the pandemic meant developing a closer relationship with nature. We all felt isolated, and as an immigrant, I knew that feeling particularly well. Quarantine isolation added to immigration isolation.

So, I started working on my mental health. I focused on self-care and tried to overcome isolation and homesickness by developing a relationship with the surrounding natural environment.

In search of a lost river, my journey began with walking. It was not a simple walk. It was walking with the camera and actively looking and listening for traces of the lost river. It was also capturing, through multimedia, the spaces built on top of it. I walked along the river with all my senses engaged. I was in search of both the lost river and a sense of belonging to my environment after immigration.

Therefore, this walking process served as my own form of self-care. I followed the stream through the woods and walkways. The year long. All my senses attuned to both the visible and the invisible stream below me. When I walked the route of Taddle Creek, any natural undulation of the landscape served to remind me of the river underground. I walked while aiming to fully absorb my surroundings in the quest for a vanished creek.

And as I walked, I wondered: Was I also in search of my wandering soul and my lost roots? As a new immigrant, was I not also looking to find a connection to my new homeland, dealing with my own loss of place? After all, as Hafez, the most influential Persian poet of the fourteenth century, notes, the flow of water and the flow of life move in parallel.

Buried Rivers of Toronto

Like elsewhere in the world, many rivers in Toronto have been buried throughout the history of this city, mainly in the eighteenth and nineteenth centuries. These rivers were always an important part of the city's landscape, but with unsustainable urban developments and increasing waste management challenges, they were eventually incorporated into the urban sewer systems.

Taddle Creek, one of these hidden rivers, is close to my home. This proximity made a long-term, process-based graduate experiment more feasible. In addition, because there are very few traces of Taddle Creek aboveground, I felt that it represented an important challenge to represent and celebrate the creek through my master of fine art project. My sense, in this case, was that the concept of absence is more profound than what is revealed. Although many similar streams have been successfully daylighted, this has not been the case in Toronto.

Taddle Creek starts at the Wychwood Barns area, just above Davenport Road, very much part of the core of the city of Toronto. Originally, the creek flowed through what is now the University of Toronto, into Toronto Harbor, near the old Distillery District. The only remaining indication of the creek aboveground is found in Wychwood Pond, although a small park is named after the creek and supports a large sculpture of a jug. *The Vessel* is a water fountain that is sculpted into the form of an almost-six-meter-high jug, utilizing miles of stainless-steel tubing and reflecting the approximate distance that the creek flowed from the park to Lake Ontario.

Another significant trace of the river is found on the grounds of the University of Toronto, in an area called Philosopher's Walk. The ravinelike topography serves as a reminder of where the creek flowed when it was aboveground. Interestingly, on occasion, the flowing water can still be heard along the walkway as it winds its way through the sewers underfoot.

Similarly, on central university grounds, the river's trace is felt through the topography in front of Hart House, where McCaul's Pond used to lie. As the historian Chris Bateman states: "The water would have sat almost exactly on the site of the grass in the middle of the circle. The irregular shape on the west side of Queen's Park Crescent is probably the best evidence on today's street grid of Taddle Creek's existence."[1]

But mostly, Taddle Creek remains invisible, buried beneath the roads and walkways, gurgling through residential basements from time to time, but most often, absent from view and from residents' awareness.

Not only is the creek invisible; so too is the human heritage— the people, culture, language, ways of life, and attitudes directed toward the earlier landscape and waterway. One could argue that the same colonial attitude that made Indigenous people invisible to settlers' eyes also allowed their land and water to be polluted, eventually buried, and thus rendered invisible.

An important part of my multimedia graduate project included making a video titled *Taddle Creek: Flowing to Parliament*.[2] In an interview in that film, Jon Johnson, an expert on precolonial Toronto history, supported this argument about the link between human and environmental heritage:

> The timing of the burying of the river and the timing of the erasure of Indigenous presence on the land are parallel. They happen, more or less, at the same time. The river was, of course, an incredibly important part of the landscape for the Mississaugas, as there were... important fishing spots

seasonally... and I imagine that there was a trail that went along Taddle Creek... [and] that it was a really well-traveled route. There was a lot of activity there. And then, the pushing of Indigenous peoples off of their land in Toronto, which happened around the early 1800s, was followed pretty quickly by the burying of the river in the 1800s. Thus, there is a disrespect for other peoples but there is also a disrespect for the waters, that we see in the way that the waters and the people were treated in this area of Toronto.[3]

The values behind polluting rivers through the dumping of sewage and eventually burying them were completely different from those of First Nations communities toward nature and water. As Jon Johnson describes:

There is a common understanding in a lot of different Indigenous cosmologies and thinking around water as a lifeblood of Mother Earth. The same as the blood in our body, waters carry around messages and nutrients and important things that we need to survive. That is how a lot of First Nations people conceptualize water, and of course, we are... part of a whole interconnected web of interrelationality, and we have to conceptualize ourselves as not separate from the earth, but rather, we are still in Mother Earth's womb.... Therefore, everything that goes into that mother's blood goes into us and affects us directly.[4]

This philosophy toward rivers is very different from the settlers' approach, which was more utilitarian and narrowly human centered. Perhaps Indigenous concepts about water teach us where we went wrong. What we need in order to respect water in the city is a change in value frameworks rather than just another type of technology. Engineering the environment through attitudes driven by human-centered, utilitarian interests has proved destructive. We have much yet to learn from the value accorded to all creatures and natural elements—including water—by Indigenous peoples.

My Project: Engaging the Invisible

I originally learned about the buried rivers in Toronto through the Lost Rivers website.[5] Because I lived downtown, I had been reflecting on how different the sense of place might have been in these urban spaces had the hidden rivers remained visible. Scientists know about the considerable environmental and ecological impacts of burying rivers. But I was more interested in exploring the broader issue of how humans have reshaped the natural environment of urban areas. I hoped to improve our understanding of the city as part of nature rather than apart from it. Ultimately, my project was a way for me to appreciate and cherish both Toronto's natural and cultural heritage.

Inspired by Sarah Pink's idea of walking with video, I decided to take an experimental approach to documenting the course of the creek.[6] In the words of Trevor Paglen, an artist and geographer, "Experimental geography, as it sounds, is more experiment than answer."[7] In his view, "Experimental geography should be considered as a new lens to interpret a growing body of culturally inspired work that deals with human interaction with the land."[8]

For my project, I was determined to explore innovative and inspiring ways to represent an invisible river that has few traces aboveground. The experimental and practice-based qualities of walking with a camera helped me discover what I would not normally notice in my everyday walks. Specific impressions about the urban fabric emerged—ideas that reveal themselves only in long-term experiments at different hours of the day, in various weather conditions, and in different seasons.

Walking with video, as Pink defines it, is "a phenomenological research method that attends to sensorial elements of human experience and place-making."[9] As she writes, "After a few times going to a garden and walking with the camera, the garden was no longer simply an imagined place but a physical environment in which many memories and meanings were already invested."[10] What she describes is the process by which a space gains meaning and turns into a valued place.

On the strength of such an embodied approach to the creek, I decided that my research would be both process based and practice based, drawing from "sensory ethnography" as a new method of discovering landscapes.[11] *Sensory ethnography* can be defined as "an approach to doing ethnography that takes account of sensory experience, sensory perception, and sensory categories that we use when we talk about our experiences and our everyday life."[12] It includes sensory mapping or what is sometimes called sensory geography.[13]

Each of these terms, overlapping in meaning, reflected my desire to avoid abstract categories and instead investigate a tactile, embodied trace of a hidden stream while mapping out my own loss of place—and search for meaning—in an unfamiliar Canadian environment.

My method was based on four interweaving types of sensory inquiry: walking, looking, listening, and touching. I began by walking with a camera. The camera focused my attention on elements that I had previously taken for granted. It framed details of my walks—sights, sounds, smells, and feelings—that I would otherwise not notice. As Hamish Fulton and other walking artists know, the act of walking can be where art happens. Fulton documents his walks through "walk-texts" and photographs, through records, always in his own words, emphasizing the importance of the human experience.[14]

For me, walking shaped my experience of urban place and of Taddle Creek. When I walked, I saw traces of the river. Even when I could not see any visible evidence of the water, I found myself imagining a stream underfoot. I was able to hear the rushing water in my mind. Being conscious of the buried river through these many walks has added a new dimension to my image of the city and of my new home.

Interestingly, walking as an embodied, careful experience made other senses active and more receptive—senses such as active looking, listening, and touching. The problem with much urban development today is that planners take a bird's-eye view of the city—an omnipotent, overhead view through aerial photographs, satellite images, and abstract maps. Walking invites a very different engagement with place—one that reveals the embodied human perspective, with all its attendant values, cultural worldviews, and aspirations. Planners could learn much from sensory geography and an embodied sense of place when it comes to designing with water.

During these walks, documenting sounds was particularly important. For this reason, my film depicts the existing route of the underground creek together with a spectrum of river sounds. Jon Johnson's voice augments these images through his own reflections about the histories of lost rivers, Indigenous people, and their ways of life—histories that, like the rivers, are so often hidden from settlers' eyes.

The sounds of my footsteps on different surfaces, combined with the images of the map or the videos of me traveling the creek's path, reflect the long process of active walking in my work. My aim was to demonstrate, as much as possible, the realistic, embodied sounds that defined my journey. I agree with the director Lucien Castaing-Taylor, who often privileges sound over the visual; in his view, "Sound is also less coded and reducible to putative meaning than picture, more evocative, imaginative, and abstract."[15] Sound is vital in my project, as it facilitates the process of ensuring that "invisibility" is heard.

My hope was to find a way to depict not the river, but the absence of it and the loss of it as a piece of nature—to invite a sense of longing for the river and for the sense of place it could create in our urban spaces. In that respect, I was challenged to represent a hidden phenomenon without directly, physically depicting it—capturing the traces (or index) of it in many ways, including through cyanotypes. *Cyanotype* comes from the Ancient Greek, κυάνεος, or *kuáneos,* meaning "dark blue," and τύπος, or *túpos,* meaning "mark," "impression," or "type." It is a slow photographic printing process that produces a cyan print often used by artists exploring monochrome imagery (figs. 1–2).

The cyanotype has been an important technique used in nature photography, and it is frequently used by botanists.[16] I made cyanotypes that are eco-friendly and experimental. The traces were never completely under my control: the water controlled them. Consequently, they have free, organic, and diverse shapes. This method gives agency to water as a nonhuman or more-than-human natural element to make relatively random shapes.

The images are contact prints, emerging as an element of a process in which I placed part of the cyanotype-coated paper between two pieces of glass and then put them into the water of Wychwood Pond. In this way, the water washed part of the cyanotype chemical, and a watercolor-like mark of water remained on the cyanotype after being developed. The images were eventually displayed as part of a public exhibit.

My documentary about Taddle Creek included other important visuals, such as a sequence of photographs and videos of what is currently visible aboveground and a map that shows the locations I walked, as well as the course of the river before it was buried.[17]

The video begins with the natural flows of leaves moving in the wind. As I move south, farther into the downtown core, the more mechanical the flow appears: I show crowded escalators in the downtown mall—the Eaton Center—and cars moving in an intersection where the river used to flow, today replaced by a mass of auto sales lots. In the end, the route of the waterway reveals a cross section of the city, with its different rhythms and meanings.

Finally, the documentary includes interviews with Helen Mills, founder and codirector of Lost Rivers Toronto. For more than two

Figure 1. The process of creating the cyanotypes

Figure 2. Documenting the trace of water on the bottom of the
watercolor paper.

decades, Helen has led group walks along many lost streams. The other person interviewed is Jon Johnson, an Indigenous person with a deep knowledge of the precolonial history of Toronto. He provides an important perspective on the Indigenous context of our lost Taddle Creek. I also included poetry about lost rivers by an Indigenous artist, Leanne B. Simpson, a Mississauga Nishnaabeg writer, musician, and academic.[18] Simpson's poem "Life by Water" is about her experience, as an Indigenous person, walking along Toronto rivers. In this poem, she calls all the missing and murdered species that once lived in the river to walk with her, naming them by their Indigenous names.

Concluding Thoughts

The primary elemental quality of water is often forgotten in cities, particularly when waterways recede underground and are absent from everyday experience. Recovering memories of hidden streams like Taddle Creek through walking with video helped me to acknowledge and bear witness to both the historical and natural sources that continue to sustain our urban environments.

My documentary, *Walking with the Invisible*, encourages looking and listening to what cannot be immediately seen or heard—a lost piece of nature in the city, retrieved and represented through an installation of video, cyanotypes, and sound. Certainly, along the length of the river, there are a few spots where one can physically hear the river at certain times of the year, and I recorded those sounds in my film. That said, most of the time, from above the ground, the water remains buried and silent. Accordingly, the project serves as a reminder that the impact of water on our lives remains present, even when physically absent from view. For me personally, it serves as a reminder of how I am also slowly becoming better acquainted with my new home in all its complexity.

As you walk through your own city, I recommend that you ask yourself: Where are the waterways? And how do they continue to sustain the forces that make the city a living place of memory and possibility?

notes

1. Chris Bateman, "The History of the Long Lost Taddle Creek in Toronto," *blogTO*, August 5, 2020, https://www.blogto.com/city/2012/03/a_brief_history_of_taddle_creek_torontos_lost_treasure/.
2. The video is available at my website: https://www.marziehmiri.com/walking-with-the-invisible.
3. Jon Johnson, interview with Marzieh M. Miri, Toronto, 2021.
4. Johnson interview.
5. The Lost Rivers website is available at http://lostrivers.ca/disappearing.html.
6. Sarah Pink, "Walking with Video," *Visual Studies* 22, no. 3 (2007): 240–52.
7. Nato Thompson, "Geography as Art, Art as Geography," in *Experimental Geography: Radical Approaches to Landscape, Cartography, and Urbanism,* ed. Nato Thompson and Independent Curators International (Brooklyn, NY: Melville House, 2009), 13–25.
8. Thompson, "Geography as Art."
9. Pink, "Walking with Video," 240.
10. Pink, 243.
11. Sarah Pink, *Doing Sensory Ethnography* (London: Sage Publications, 2009).
12. Sarah Pink, *What Is Sensory Ethnography?* (London: Sage Publications, 2011), https://doi.org/10.4135/9781412995566.
13. Originally, I came up with the term *sensory geography* on my own. But later I realized that others were using it and drawing on its ideas as well. See, e.g., Nina J. Morris, "Teaching Sensory Geographies in Practice: Transforming Students' Awareness and Understanding through Playful Experimentation," *Journal of Geography in Higher Education* (May 2020): 550–68, https://doi.org/10.1080/03098265.2020.1771685.
14. Alan Macpherson, "Sensuous Singularity: Hamish Fulton's Cairngorm Walk-Texts," *Critical Survey* 29, no. 1 (Spring 2017): 12–32.
15. Scott MacDonald, *Avant-Doc: Intersections of Documentary and Avant-Garde Cinema* (New York: Oxford University Press, 2015), 409.
16. A great example of this technique can be found in Anna Atkins's extensive photobook of botanical specimens made through the cyanotype process, *Photographs of British Algae: Cyanotype Impressions,* 1843, https://publicdomainreview.org/collection/cyanotypes-of-british-algae-by-anna-atkins-1843/.
17. Walking with the Invisible, https://www.marziehmiri.com/walking-with-the-invisible.
18. Leanne Betasamosake Simpson, "Life by Water," in *An Enduring Wilderness: Toronto's Natural Parklands,* ed. Robert Burley (Montreal: ECW Press, 2017), 102.

Learning to Swim

Elizabeth Bradfield

—after Bob Hicok & Aracelis Girmay

Now forty-five, having outlasted some of
myself, I must reflect: what if I hadn't been held
by my mom in the YWCA basement
pool, her white hands slick under

my almost-toddler armpits, her thumbs
and fingers firm around my ribs (which
is to say lungs), held gently as a liverwurst
sandwich and pulled, kindly, under?

What if I hadn't been taught to trust
water might safely erase me those years
I longed to erase or at least abandon care of
my disoriented, disdained body? I might have

drowned instead of just ebbed, never slid
from given embankments into this other
natural course.

                  Drift and abundance in what
she offered. The wider, indifferent ocean
of trade and dark passage not yet

mine to reckon. And so now, sharp tang
of other waters known, I am afloat, skin-
chilled, core-warm, aware of what lurks
and grateful to trust and delight
in our improbable buoyancy.

The Crawdad Quadrille

Margo Farnsworth

Will you, won't you, will you, won't you,
Will you join the dance?
—Lewis Carroll, *Alice's Adventures in Wonderland*

C rawdads—that's what we called them.

My first memory of dancing with rivers is the smell of sun-drenched mossy rocks harboring soggy, half-sunken snags, the green of willows, and maybe a drying dead fish at the waterline. I remember Dad crunching across the gravel and me, all of five years old, traipsing after him to see what would come next. He waded out into the water. I followed.

"Look at this." He pointed at what looked to be a miniature lobster, like those I had seen in the grocery store tank. Water flowed around his calves as he instructed me on the finer points of crawdad hunting.

"Will they pinch me?"

"Maybe so, but it doesn't hurt much. See there how they scoot backwards? The trick is to get one hand behind them while you gently wave the fingers of your other hand where they can see—kind of like decoys. They'll scoot away from that one, right into the hand behind them."

It seemed like kind of a dirty trick to play on the crawdad, but, if I put them right back in the river, I thought it would be OK. So, for the next hour or so, while the adults were alternately preparing canoes and laughing at whatever adults laugh at,

I was engrossed in perfecting the art of crawdad hunting in an Ozark stream.

I tested techniques in water up to my thighs where the current asked me to think about swimming, in knee-deep water where I could contemplate leaves dancing in the current along with nearby birdsong, and ultimately in water a little over ankle-deep where the hunting seemed best.

There were so many crawdads waltzing there; big ones, little ones, some with speckles on their claws, orange eyeliner, pinchers held like barbecue tongs until they nipped at my young flesh. *Ouch! Dang!* And *plop!*—back into the flow they went. We were high up on the river, where the water ran swift and cold with rivulets running in from here and there to enlarge the flow and add nutrients from beyond my sight line. Minnow schools interrupted my stalking with their mainstream movements, shimmering upstream then immediately reversing course. Darters dashed in the same pools where the crawdads coasted then rested, coasted then rested. In the line of air above the river, a slate blue and white kingfisher rattled swiftly downstream. All of us creatures shared the world and water that morning.

Like crawdads, humans are a limbed sack of organs and moving fluids. In this, all living organisms are basically the same. Yet within species, and among human beings, each one manifests itself in unique ways. Water is much the same—not in terms of organs, of course—but in its ultimate similarities among its various forms, yet with variations among them all. Although its structure is reliably two atoms of hydrogen and one of oxygen, the truth of water is determined by its surroundings and location. Each body offers something distinctive—collections of seeps flowing to creeks then lakes, rivers meeting oceans at estuaries, tidal pools sidling up to seagrass beds next to prodigious coral walls dropping into the dark. Some bodies flow while others seem still. Some are rich with nutrients while others carry soluble arsenic, uranium, and the like. No matter their place or payload, the water bodies each constantly cycle through flow, evaporation, and then rain, snow, hail, or some other form of atmospheric liquid, solid, or gas. Frozen glaciers, desert potholes—all so different and yet each one has those atoms in common. They also share relations to the air and land portions of Earth. Generally, we think of water's relationship to land as nurturing, bringing forth green and fruits and shade. But as with any relationship, water can overpower and inundate, strip out foundations or entire towns, reduce the fecund to snarls downstream, bury the lot or simply leave cold, sharp skeletons of what once was—much the same as humans do.

Perversely and thankfully, upheavals can also be beneficial when applied by nature's clock and rules. Floods can bestow the kindness of new soil. Snow can insulate young plants and hibernating creatures. Healthy cycles yield a verdant sum.

These water worlds harbor tenants from crawdads and shiners to great salmon, bull kelp, and orcas. They deliver their bounty to lakes that pool and flip with the seasons—cold and warm water trading places regularly. Along with aquifers belowground, lakes store this precious substance and support other lives at their shores. Otters, coyotes, and trees drink, commune, and create new generations within that realm. Estuarian nurseries and oceans, with all their own habits and cycles, hold more neighbors who dance with each other to mutual benefit. Humans come and go as well through each of these bodies, taking what they want and only sometimes, in infinitesimal ways, giving back. Water gives, takes, and renews in a constant regenerative waltz.

As water flows spreading life, freezes to aid in preserving food, or—forced to steam to turn machinery—produces electricity and other human-serving actions, this wondrous elemental cannot be continually treated in an offhand manner without consequences.

When embraced and cared for, it can foster development and growth. It can absorb death and start the cycle again, giving life to the planet. But unlike with a dance partner who might simply grimace as I step on his toe, our dance with water is governed by rules. I think of it as I buy clothes, eat dinner, walk on a sidewalk. Am I dancing too wildly with ignorance or lack of care for the amounts and quality of the water I come in contact with—the H_2O supporting all life? After all, as Mr. Carroll's lobsters demonstrate in their quadrille, it is a stately dance with specific steps and requirements. Water, in all its forms, has its needs as well.

"Please, don't take too much," it whispers, then bellows as aquifers drop. "I cannot continue to absorb and dilute what you're putting into me," it moans, as endocrine disrupters make their way into our homes.

Water does demand it not be ignored. It is, in fact, one of Earth's limiting factors. Like gravity, it has its absolutes. And yet humans, in our recalcitrant insistence on seniority, keep pushing, pushing, pushing water and one another to our flooding points, our drought points, and our breaking points. We're tripping across the dance floor.

Over time I graduated from catching crawdads inland to full immersion in underwater worlds as I learned to snorkel. Cruising along at the surface, I swam near schools of French grunts with their decadent yellow-veined bodies, official-looking sergeant majors dressed in stunning stripes, and the beautiful blue tangs moving in synchrony and inspiring artists at Disney who painted them into history as Dory darted onto our screens.

Every time I entered the water, my breath came fast as I scanned the surrounding area for sharks. Dad had taught us not to fear them, but my preteen brain evoked a cautious perspective: *Keep an eye out for the species who looked to be more bad-tempered.*

That way there's always snorkeling tomorrow. I could have saved my circumspection, as they took little notice of me.

In the meantime, there were gorgeous shells bejeweled with hard mother-of-pearl interiors to retrieve from the bottom. When I dove, there was seaweed to stare up through at the surface, floating by merit of its many air bladders. I reveled in all of this and the array of other plants and animals simply waiting for a new friend to discover them.

One afternoon as I was swimming up to the boat, I spotted a very large, very dark form in the distance. I swam a little faster. Even though I could tell it wasn't a shark (the largest thing I'd met until then), I grew more tense as I realized it was larger across than any shark or anything I'd ever seen in this watery world. Black form growing ever larger, it seemed to be flying through the water. I was almost to the ladder. Salt water seeped into my mouth as I involuntarily gasped. It was huge—half the length of the forty-foot boat. Faster. Closer. Almost to me with its wings and gaping maw. I couldn't get flippers up the ladder and had no time to discard them, so resorted to a desperate pull-up over the gunnels as the thing banked right and circled back.

In full terror I croaked, "I thought rays were small!"

Dad chortled. He actually laughed. "That *thing* is a manta ray and the only thing she's interested in eating are krill and other tiny critters. She doesn't even have teeth. She's just curious about what you are."

If she was curious, I was more so, but happier on the boat watching her repeatedly soar underneath us, filter feeding on whatever goodies seemed abundant there.

Years later, that curiosity—of streams with my crawdad friends to oceans full of creatures prospering from a vast collection of strategies they wield to move, eat, shelter, migrate, and reproduce—carried me to the doorstep of biomimicry. After reading Janine Benyus's book *Biomimicry: Innovation Inspired by Nature*, I thought everyone must already be using this commonsense methodology for design and invention. I found instead that there were still very few of us starting to explore the practice of biomimicry by observing nature, analyzing organism strategies for performing functions in sync with their surroundings, and then emulating the forms, processes, and systems that allow them to thrive.

I dove into the studies, heading straight for the source, where I learned from Janine and an inventive array of early adopters. These people respected water as an element critical to our survival but also as a substance cradling the life forms from whom we could and can learn.

Those schools of fish with whom I swam are geniuses at optimizing energy savings. By swimming in schools, fish as small as blue tangs or as large as the great albacore tuna save energy by drafting off the fish in front of them. With every wave of their vertical tails, they produce vortices that create an easier swim for the fish behind them.

Scientists at Caltech's Field Laboratory for Optimized Wind Energy are studying schooling behaviors and have begun applying the strategy to positioning vertical wind turbines. By arranging these turbines in "schools," they have improved efficiency tenfold. Once commercialized, this will elicit positive environmental and economic impacts.[1]

With biomimicry, the mother-of-pearl, of which I was so enamored, became an organism's innards from which to learn. That exceedingly smooth and lustrous material is created by organisms in both fresh and salt water by layering the mineral aragonite in an impermeable bricklike formation and mortaring it together with a composite that acts like cement. The biomimicry innovation that emulates this process can create less waste and a substance that is 60 percent stronger than steel and potentially useful in items as diverse as helmets and packaging, and in a host of other products.[2]

A relative of those greenish, floating forms of seaweed held aloft at the surface by air bladders inspired a new invention to yield energy. By mimicking the anchoring structure and leaf undulation of bull kelp found in the Pacific, the innovators at the renewable energy company BPS developed bioWAVE, an invention that sways with ocean currents, converting wave action to clean energy.[3]

The curious manta ray has inspired a team of biologists and engineers to emulate its wondrous, open-mouthed, krill-harvesting ways. Unlike many current filtration technologies, which get clogged and require high energy inputs, the manta ray's gills allow for efficient filter feeding through specialized filter lobes that ricochet food toward the throat while allowing high volumes of water to pass back out into the ray's surroundings.[4]

We are even learning from the crawdads in that Ozark stream and their larger lobster cousins. The stacked lenses in their eyes help make their often-murky surroundings brighter to them. By emulating these strategies, scientists have developed new designs for radiant heaters and handheld imaging systems.[5]

All these strategies—energy-efficient schooling, the structural strength of mother-of-pearl, energy production based on swaying bull kelp, the fantastic filtration systems of manta rays, and the reflective abilities found in a crustacean's eyes—are but a grain of sand compared to all we can learn from those who flourish in the element we call water.

Of course, learning from these organisms (and oh-so-many more) is one of the most meaningful parts of my life these days. I learn about them, then pass that knowledge to the many professionals who knock on my email, Twitter, or Instagram. But I fret about time. I've spent the better part of a lifetime on, in, or near water. Until I was an adult, it never occurred to me that our society, on the whole, was giving our waterways little thought. In fact, all of us as we dance together often claim we do not have the time or money to consider and change our impacts on the water we use or what we deposit into it. Sometimes we are simply too tired. Sometimes I am simply too tired, even as I have worked for and in water. But we are now operating on a frighteningly short rope with little time to stem the wrath wrought from a current sixth extinction. We are losing genius organisms while they are losing themselves as species. When considering this I think again of Mr. Carroll:

> Is all our life, then, but a dream?
> Seen faintly in the golden gleam
> Athwart Time's dark resistless stream?

> Bowed to the earth with bitter woe,
> Or laughing at some raree-show,
> We flutter idly to and fro.

> Man's little day in haste we spend,
> And, from its merry noontide, send
> No glance to meet the silent end.[6]

Each of our silent ends is coming. Carroll's verse suggests I should act—not flutter idly. I will continue to learn nature's ways and teach the dance steps to others wishing to do the same. And what will I teach first? I will tell my students it is time to sit patiently with nature and observe the strategies found there that address all our everyday functional challenges to move, build, and harvest energy and food, along with the host of other tasks we all face. The other-than-human elders with whom we share this world have carried on and, indeed, made good use of time throughout time to solve seemingly insurmountable problems.

It is one thing to suggest we learn from nature and emulate the brilliant strategies we discover. It is quite another to learn to dance with nature—to take time in our work, or our purchasing, to consider and honor the patterns of nature that are materially and energetically efficient, locally attuned and responsive, that use life-friendly chemistry, integrate development with growth, adapt to changing conditions, and evolve to survive.[7] But I love to dance, and in dancing with nature through biomimicry, I can honor my neighbors the crawdad and seaweed, the graceful manta ray, and teach others the steps as well.

In the end, we all dance with water and its organismal constituents. Choices we make each day, through actions or inaction, will determine whether that dance will be an orderly quadrille or the chaos of a mosh pit. Will we, won't we? If we learn from nature and mindfully practice solutions emulating other organisms, we may be fooled less by the fingers of ignorance and simple exhaustion from all there is to learn. In doing so, we may avoid backing up into our demise.

I think about this as my grandson dips his tiny toes into one of my favorite rivers. I'll teach him to dance with the crawdads.

acknowledgments

And many thanks to my dear Wildlings, who danced with me as I created this written quadrille.

notes

1. "Synergistic Windfarm Design Inspired by Schooling Fish: California Institute of Technology," AskNature, 2010, https://asknature.org/innovation/wind-farm-design-inspired-by-schooling-fish/.
2. "Impermeable Coating Inspired by Nacre: University of Connecticut," AskNature, 2017, https://asknature.org/innovation/fireproof-coating-material-inspired-by-nacre/.
3. For more information, see the BPS website, at https://bps.energy.
4. "Fine Particulate Matter Filters Inspired by Manta Rays: RICOCHET," AskNature, 2020, https://asknature.org/innovation/rticulate-matter-filters-inspired-by-manta-rays/.
5. Sherry Ritter, "How Lobster Eyes Inspired a Radiant Heater," *Make:*, November 10, 2014, https://makezine.com/article/science/energy/how-lobster-eyes-inspired-a-radiant-heater; Chris Ashcraft, "New Design Innovations from Biomimetics," *Creation*, July 2010, https://creation.com/lobster-eye-design.
6. L. Carroll, *Sylvie and Bruno* (London: Macmillan, 1889), vii.
7. Biomimicry 3.8, "Life's Principles," 2013, https://biomimicry.net/the-buzz/resources/designlens-lifes-principles/.

The Fidelity of Water

Geffrey Davis

I.

Thousands of miles from home, you wake
in a cheap hotel with thirst so urgent you have
no choice but to find the bathroom faucet
with your mouth, drink deep, and understand
the daily sigh made by bodies everywhere
in this small town. Your new cushy job gives you

bottled artesian water, which you consider
as you taste the tap. Used to be there was
no good distance between this rivery tang
and your fluid desires. Used to be you'd shove
aside a sweaty friend or jab a sibling
for the first shot at placing lips to the only

neighborhood source. Used to be no future
in yearning.   Feel how far you've come?

II.

Today a flood, and you see the risk
in proximity, in life stretched by loving

both a river and the rain:—to watch what feeds you
run dangerous, the Biblical possibility

of nurture rising into a final rage.
Father rain. Mother river.   When it pours,

and you love when it pours, this river turns
tannic with a turbulence you can recognize

as home—: Mother rain. Father river.
But what course isn't threatened when

the right season licks its lips? What epic
confluence can avoid the violence

of giving and taking such shape? Mother father.
River rain.   If given the chance, you too will empty

or swell—will lay claim to every unrooted thing
in the name of a love you learned from flood.

Thirst: A River's Marginalia

Lyanda Fern Lynn Haupt

W here to begin this story—with the bird, the book, the lake, or the river? Eventually they will all become entangled, so I suppose it doesn't really matter. I'll start with the river.

The Elwha River is a central life bearer within the Olympic Peninsula ecosystem in my home state of Washington. From time immemorial, the Lower Elwha Klallam people were sustained and nourished by a flourishing population of regal Chinook salmon, who swam and played and spawned in the river's waters.

But from the early 1900s, and until very recently, the Elwha has been held captive by two dams, one low and near to the ocean, one high and near to the mountain glaciers. Lower Elwha Klallam tribal member and fisheries worker Mel Elofson recalls his grandmother, Louisa Sampson, sharing stories of the Chinook salmon here, "so thick a person could walk from one bank to the other across their backs." The dams were completed in 1911 and 1926, immense concrete barriers without fish passages; over the course of a century, native salmon and anadromous trout have disappeared from 90 percent of the Elwha's watershed.[1]

The ecological devastation that these dams wrought upon the salmon and the forests and the Klallam people and all the shared life of this inimitable temperate rainforest system is beyond reckoning. And yet there is no getting around an ironic truth: the lake created by the second, highest, and largest of the dams—Glines Canyon Dam—was very beautiful.

Lake Mills came to look natural over the decades, with firs, redcedars, ferns, and mosses reaching toward a curving shoreline. There evolved a lakeside trail beyond which the waters of the Upper Elwha flowed, rushing toward the lake with feral abandon, unaware of the violent barriers about to halt them.

The enchanted mountains and waters above the pretty pretend lake were the stomping grounds of my formative years. I backpacked the trials, carried notebooks and whittle-pointed pencils, composed my master's thesis on radical environmental activism by dappled forest sun, by river-glisten, by tent-hung headlamp. Lake Mills was the starting and ending point for my rambles into the higher, wilder hills, so I walked the lake edge often.

The rocky convergence of river-into-lake was frequented by my favorite bird. I cherish American dippers in part for their association with beloved places—turbulent stone-strewn rivers and streams in mountain forests. But dippers are also as cute as pie. Round, slate gray, wide eyed. And although it would be a mistake to describe these wild river birds as tame, I will say that dippers in general are unwary of human presence. They have the advantage, after all, of being able to flap to a stone surrounded by deep rapids in a fraction of a second. And so we humans are allowed to draw near to the miracle of dipper life.

Dippers are passerines—songbirds—and among this large order which includes everything from chickadees to swallows to ravens, they are the only species so extravagantly adapted to riverine existence. The dipper's song is water inspirited: a torrent of trills and warbles, high-pitched to carry above the river's own clamoring song. They plunge their heads fearlessly into rushing water in search of prey. "Trout with feathers" they are called by some, as they haunt the same pools as trout, feather-scales shining, eating just what a trout eats: invertebrates of any kind, damselfly nymphs, small and round fish eggs. They are merbirds, closing their specialized nostril flaps to keep from drowning when underwater and possessing hyperoxygenated blood. Like many birds, they have air sacs that assist in flight, but they can also compress those sacs to keep from being buoyant during their underwater forays. They seem not to notice in the least the dangerous swiftness of their homewaters. When they spot something potentially delicious, they fling their entire little two-ounce bodies into the fastest flowing water on earth, wherein they not only swim but also often walk—*walk*—upon the river bottom. Small, feathered Poseidons.

I marvel. In six inches of this water, I would stumble; up to my knees I would fall; any deeper I would be swept away, and my mother's ever-present worry of finding my cold, dead body against some mountain glacial erratic might just come to pass. So when stopping here for lunch, I would climb gingerly to a wide, safe boulder, deploy my little picnic, and bask with delight among the dippers.

During this time, I was obsessively reading the work of Emily Dickinson. I'd carried a volume of her complete poems with me everywhere since I was a junior in high school, and had filled the margins with diary-like notes in pencil and colored pen, and on scribbled Post-it notes. There were youthful, now-embarrassing comparisons of the poems to my own thoughts, my waking dreams, my night dreams, the moments of my days. There were cross-references between Dickinson's poems and those of other writers. There was dirt, spilled coffee, and a little blood. The pages were filled with the ephemera of life, as the books dear to us will be—leaves, feathers, love letters. I carried it to college, to Japan, to the remote tropical Pacific. I carried it everywhere. And although I weighed every ounce of my food and clothes when packing for a backwoods trip, still I carried, always, and surely a bit foolishly, this volume of Emily Dickinson's poems that weighed more than my sleeping bag.

One sunny day at the convergence, I scrabbled to my favorite lunching boulder and unfurled my minicamp: Wheat Thins, apple, notebook, poetry. What happened next is a complete blur—replayed a million times in my desperate memory with no sense of exactly how things unfolded. I must have knocked the book with my toe? Too fast for me to do anything more than gasp, my precious

Emily tumbled into the water and *whoooooshed* downstream. The water was sweet green and glacier-clear. I watched the pages flap beneath the current before catching against a stone, words visible but artistically obscured by the sparkling ripples above.

I had no way of getting to the book. I was just eight or ten feet away—so close!—but the book was in a shallow protected on all sides by a moat of deep rapids. I spent the next half an hour trying to pick a path to the book via other stones, and another hour attempting to use fallen branches to dislodge the book until finally realizing it was no use; I would never get it back. I returned to my lunch boulder and cried.

Oh, but then—through tears I watched as one of the two dippers on the river that day hopped onto a stone near the book rock and then *onto* the book rock. She bobbed up and down the way dippers do. While I stifled my sobs and held my breath, the bird hurled herself into the water, dove beneath the current, and stood on the river bottom as if in a still breeze. Then it happened—the dipper *noticed* the book. She faced Emily Dickinson's poems, paused a moment as if to take in the words on the page, walked a few steps, let the torrent catch her, then emerged—glistening—on another boulder nearby. I sat still and breathless. In what unique dipper way did she take in the words, illumined by water beneath her? *Only a bird will wonder...*

I will not impose upon the bird, but in my imagination, I knew what was true for me. There, the Elwha dippers and Emily's poems became fluent with each other. And why would they not? For my imagination knew, too, the page the book was open to, and the poem upon it:

Water, is taught by thirst....
Birds, by the Snow.

The deepest appreciation of water can be formed only through the crucible of thirst, yes. But beyond the literal, what is this thirst?

Emily Dickinson was a student of philology and etymology, always seeking depth and vitality through word choice that would bring her spare use of language to startling fullness. Language lives and evolves in much the same way organisms do: by genetic mutation (when an unpredictable new attribute catches on); by geographical isolation (as when birds or plant seeds cross water, do not interact with their mainland counterparts for some generations, and become biologically endemic to their place); and by practical usefulness (in a neo-Lamarckian fashion). A word is as alive as a bird.

The evolutionary insight of etymology is often astounding, but the pathway to the word *thirst* took my breath away. It happens that thirst itself is a river-word, from the Latin *torrere*, a violent stream, a rush so wild that it leaves a place dry in the aftermath. A *torrent*. I think of the Glines Canyon Dam, greedy, holding water back only to release it in an unnatural, torrential rush. Sediment, nutrients, fish runs, bank margins, all confused and lost and wondering. An entire ecosystem parched and thirsting.

I don't remember many of the specifics from my youthful marginalia, but I remember my unanswered query in pink colored pencil in the margins alongside the dipper's poem: *For what do I thirst?* What emptiness, what suffering, what longing, what torrent brings me to the river? Brings any of us?

During the writing of this piece, I opened my notebook to a clean page and scrawled across the top margin my long-ago question, no less relevant in this time of life, this earthen moment: *For what do I thirst?* I kept the notebook open on my desk, carried it in my rucksack during my daily wanders. Answers emerged—true answers perhaps, but flimsy, thin, obvious, silly. I penciled them uncertainly onto the page: *Beauty. Connection. Peace. Chardonnay.*

Hmm. I set *thirst* aside for a bit to focus on... *Birds, by the Snow.* Dickinson was attentive to the natural world; outside her Amherst window, the seasonal cycles guided the migratory birds' flight—orioles, robins, sparrows, warblers—the coming of water in the form of winter snows spurring their move to a more life-sustaining clime.

The Elwha dippers know this snow-water in their own way, as their river is sourced and fed by two main glaciers settled high in the Olympic Mountains. In the 1890s, this Klallam land was mapped by the white surveyor Theodore Rixon, who named one of the ice fields Carrie Glacier, for his fiancée, Caroline Jones.[2] Carrie Jones-Rixon was a near contemporary of Dickinson, their lives overlapping temporally from opposite coasts. When Carrie Glacier was mapped just ten years after Emily died, Carrie Jones was romping the Olympic mountain trails in her long skirts and bustle, fishing the Elwha, painting the landscape in oils, and teaching art, poetry, and music to children.

Soon after the first dam was completed on the Elwha, the anadromous Chinook salmon returned from their years of growth and life in the Pacific to spawn in their natal river. Enormous, rigorous, shining, ready. When met by the wall of concrete, they rebelled—slamming their bodies into the dam again and again until they were ragged and bloody. They did not give up, but died of injury, exhaustion, and illness from too-warm water. Louisa Sampson watched from her homestead along the river, and Mel Elofson recorded his grandmother's memories: *She said she was devastated to see so many thousands of salmon dying at the base of the dam; because there was no other passage for them to move further upstream many went unspawned and she said she was quite distressed from this experience. All she could do was go back and (she) said she cried for hours.*[3]

But the dams could not quieten the spirit of the river. Salmon ghosts continued to swim among stones, their swish and slosh keeping the song of the Elwha alive. The timbre and tone of the waters beckoned tribes, activists, researchers, hikers, tourists, and everyday humans within and beyond the watershed to remember what they had always known: that a river is alive only insofar as it flows in dynamic relationship to all the beings within the watershed. The elemental connection between glacier-river-ocean and the salty stream of our own blood flow inspired recognition of a power much deeper than the trickle of hydroelectricity provided by the dams.

It took decades of research, activism, and public education, but finally Congress relented, and dam removal began in 2011—the largest such project in US history. Fishery scientists wondered how many years it would take for the Chinook to return. No one predicted the celebratory truth: in just six months, the first salmon spawned above the lower dam location. They dug their redds in the river's margins, laid their eggs, and died a beautiful, natural death. Their scarlet bodies began to feed all the beings of the river, including the land beings of the watershed—redcedars, spotted owls, black bears, the ghost ancestors of wolves who once roamed here, the unborn wolves who will one day return, like the salmon.

Soon after these first salmon spawned, dippers were observed foraging their eggs, another warp thread in the great weaving of a wild river's recovery. After a few years, Lower Elwha Klallam tribe biologists mist netted dippers and administered delicate blood tests, finding that with access to salmon eggs and salmon-nourished waters, dippers had become fitter, stronger, and healthier, and had higher survival rates. Instead of expending valuable energy to forage widely, they could stay close to home, singing to one another upon their favorite Elwha River stones, day after day, year after year.

It is so clear now.

This is the thirst. To *continue*. The word itself is a poem, evolving through Latin roots that suggest a flowing "container," a process that "holds together in fullness." My own deepest thirst twines with the dream of these birds, these fish, this river, this forest, this earth. We thirst to continue, to simply go on. Not to live forever, but to take refuge in the sweet cycle. To go forth, to return, to know sustenance, to offer back in fullness through the nourishment of our own lives and bodies. Again, again, again.

Carrie Glacier is receding now. Climatologists predict that the ice field will be gone by 2070, along with all the Olympic range glaciers. Snow and rain will fill the ecosystem's tributaries, but the glaciers serve as icy reservoirs, slowly feeding the rivers with the cold waters that salmon require. The Elwha will surely slow and warm. No one knows what will happen, who will continue, who will not. The river thirsts even as the freed waters run.

Where does this leave us, we who would honor the waters? Science offers little hope that the radical reversal in course required to halt, or even slow, the loss of the glaciers is possible. Tearing down walls of concrete is a child's game next to stemming glacial melt. Yet when I sit in riverine reverie these days, my bare feet in the Elwha's cold current, I turn my face toward the mountains where the salmon are rising, up and up like their ancestors to the highest mountain streams, their bodies still salty with ocean brine. I know that we must rise with them.

Emily Dickinson never saw the sea, but she sensed that the ocean's waters could not be separated from her self, her life, her gifts, her pen. Where does the glacier become the river, the ocean, the rain, the poet's liquid ink, the bird's eye-glisten, the sweat tickling the back of my neck? The river song answers: nowhere, everywhere. In the depth of our elemental interconnectedness we are called, with every cup of clear water we lift to our thirsting lips, to live in simplicity, in gratitude, in renewal, in praise, in protest.

For a while, I mourned the loss of my book, but I was consoled by the fact that I finally had an excuse to spend my meager student stipend on the Belknap Press's newer *Poems of Emily Dickinson*—the edition of my dreams. More than the book itself, I lamented the lost marginalia, the comforting mycelia of my small life—often at first, less and less as the years went by. But whenever I think of the Glines Canyon Dam coming down, I sense that none of it was lost at all. All is gathered in the torrents. The freedom of the river. The seeing of the dippers. The return of the salmon. The presence of the poet. The scribbled details of our tiny lives. All of this, all of us, mingled—imperfect, sacred, and whole—within earth's ever-flowing, ever-thirsting waters.

notes

1. Lynda Mapes, *Elwha: A River Reborn* (Seattle: Mountaineers Books, 2013), 102.
2. Until very recently, this was one of just two ways that anything at all in the scientific world—glaciers, mountains, birds, ferns—became named for women. White male explorers or scientists would confer names that honored either rich or royal women patrons or their own wives.
3. Mapes, *Elwha*, 102.

The Rule of Thalweg

Anna Selby

My mother wades in behind me. I walk on ribcage,
riverbed, water-level no higher than my hips, current
pulling below; no, coaxing, dozy, slowed.

One New Year's my mother walked
over this frozen weir—drunk, homebound,
fast snow melting like salt on her eyelashes.

I tiptoe down
to the lowest part—thalweg, pregnancy seam,
where the river begins to be owned.

Last September, after the first heavy rains
I travelled to see the salmon amassing:
a pucker of mouths at the surface,

huge apostrophes
stirring in their spawning colours, new
bodies turning into the estuary.

October, my birthday,
I followed them back: slack tide to molten Severn,
loosed canoes bucking past me like riderless horses.

When I breach the other side
and climb out, the scorpion larvae of stoneflies
hug my thighs, their feathered edges, my feet

falling. I hold my hand out
to feel the tributaries.
My mother is 73.

This is the hottest day, the lowest river mark.
She is back on the bank,
dry bag dripping, dark prints.

*Only 2 percent of all rivers in England and Wales have public
access rights and since the banks and bottoms of non-meandered
rivers are legally private property, permission is required from
landowners to walk on the banks or bottom of those waterways. It
is classed as trespassing to put a foot down and touch the bottom
or territorial line, the thalweg.*

Are Rivers Persons?

Ingrid Leman Stefanovic

"We are born alone, and we die alone."

So declared my friend and colleague years ago, in our final undergraduate philosophy class. Extraordinarily bright—though admittedly somewhat cocky—my friend often seemed more insightful than even our professors. So, his pointed interpretations of existentialist views on death and dying duly impressed me at the time.

But then, some years later, my beautiful son was born. Eighteen months after that, my stunning little daughter arrived.

And it was then that I knew that my friend had been so wrong.

Neither of my children was born alone, as no child can be. *My own* embodiment was the condition of *their* very existence, just as much as *they* seemed to be mysterious extensions of *my own* selfhood. Certainly, none of us was ever alone through the process of birth.

Today, my friend would argue that I was missing the point. Existentialists, he would explain, meant that no one could take another's place, either in birth or in death. And that may certainly be true in a blatantly empirical sense.

But where my friend was wrong was in his solipsistic characterization of life in terms of aloneness. No one is really born alone nor, would I suggest, do they die alone. Always, we stand *in relation* to other people, places, contexts, temporal moments, and lived worlds.

I invite you, the reader, to try to imagine yourself truly alone, independent of other people, animals, living entities, environments. You will, no doubt, find this task to be impossible. We do not live in a void. Hardly a trivial realization, the experiment reveals that our very existence is defined not as lone entities in the world but rather always *in relation to* others and the environment in which we are emplaced.

This essay explores something about this relatedness—a relatedness that is complex, rich, and not always uniformly positive though certainly, ultimately, mysterious. Specifically, I want to consider how we relate to other *persons* and whether rivers and waterways can be meaningfully defined as persons as well.

This may sound like a strange idea—even ridiculous—until you realize that, already in the news, there have been reports of some rivers being accorded precisely such status of legal "personhood" in various parts of the world.[1] What does that mean, and why are some countries engaging in such innovative judicial steps? Can water truly, legitimately be described as a person, and if so, how and why would anyone feel compelled to do so? Is there some legal utility to this move, or might there be more to the story?

Let's begin by exploring where and why some people have felt that it is important to extend this notion of personhood to rivers and waterways. After reflecting briefly on the origins of this notion of personhood in Western thought, let's then ask ourselves: other than the legal utility of such a move, is there something more that we can say about the meaning and value of engaging with water through this prism of personhood? My own cagey answer to the question of whether assigning rivers personhood is a legitimate initiative will be... it depends!

Rivers Granted Legal Rights

"Each time there is a movement to confer rights onto some new 'entity,' the proposal is bound to sound odd or frightening or laughable." So wrote Christopher Stone in 1974, in his classic book, *Should Trees Have Standing*— arguably the first major work to present a legal argument to support the rights of nature.[2] In the book, he argued that, just as it took time before slaves, women, Black people, and others were formally accorded human rights in light of prior discrimination, it was finally time to afford some form of environmental rights to trees, rivers, and oceans. Stone would have been pleased to know that, in recent years, several rivers have been provided precisely such legal standing.

The fact is that rivers are in trouble. We have polluted, diverted, buried, and dammed far too many waterways worldwide. According to some estimates, only a third of the 177 largest rivers across the globe remain free flowing, and fully half of the world's major rivers are depleted or seriously polluted. In the United States, more than 50 percent of the streams and rivers may no longer be able to support aquatic life.[3] And who has not heard of the incredible case of high levels of hydrocarbon pollution causing the Cuyahoga River in Ohio to catch fire in 1969? Although such cases helped inspire the creation of the US Clean Water Act and Safe Drinking Water Act, more remains to be done to genuinely protect all our waterways. "We have to change our environmental protection in this country and across the world," says Tish O'Dell, of the Community Environmental Legal Defense Fund, "because obviously, what we're doing isn't working."[4]

Increasingly, such change involves conferring legal personhood upon the environment—specifically, rivers. How did this initiative come about?

One version of the story begins in New Zealand, where, for many decades, the Māori people engaged in land disputes around the Whanganui River. Whereas in 1840, the government had signed a treaty with the Māori tribe, granting them control over the river, that guarantee was not respected in the English version of the treaty. Finally in 2014, a resolution was reached, and in 2017, an agreement was implemented, stating that the Whanganui River "is an indivisible and living whole" and "is a legal person and has all the rights, powers, duties, and liabilities of a legal person." The new "record of understanding" includes an apology from the

Crown, adding an acknowledgment that the health and well-being of the river are "intrinsically connected with the health and well-being of the people."[5]

In fact, the Māori people see the river as their ancestor: never above nature, they see themselves as part of it. Today, one person nominated by the local tribe and another by the New Zealand federal government operate Te Pou Tupua—an office that has been created to serve as representative and guardian of the river.

In 2017, a similar initiative arose when the Indian High Court of Uttarakhand proclaimed the Ganges River and its main tributary, the Yamuna River, as legal persons. The decision would have meant that polluting or damaging the health of the rivers were the legal equivalent of harming a person. The court again appointed officials to act as legal custodians of the rivers and ordered that a management board be created within three months' time.

Sadly, in this case, the Supreme Court came to reject the High Court decision, principally because the agreement became impossible to enforce. Because the river flows through multiple jurisdictions, including the state of West Bengal, before it reaches Uttarakhand, enforcing the decision by holding Uttarakhand guardians responsible for upstream polluting was neither morally nor legally feasible. The case is a reminder that legal personhood must matter not only in principle but also in practice.

At about the same time in Colombia, a Constitutional Court decided in its ruling T-622 that the Atrato River basin would be assigned rights to "protection, conservation, maintenance and restoration."[6] Once again, government and community members were assigned the responsibility of guaranteeing the rights of the river, acknowledging the interdependence of the natural world and human communities. In 2018, the Supreme Court of Colombia similarly recognized the Amazon ecosystem as a legal subject.

Increasingly over the past few years, more and more instances of granting rivers legal rights have arisen around the world. Perhaps the stage had already been set when Ecuador and Bolivia each granted constitutional rights to the natural environment in 2008 and 2010 respectively, hoping to find a way to receive compensation from polluting companies.[7] At the same time, many seek not only financial reparations through legal channels but, rather, their actions reflect deep sacred, elemental relationships. "The river is considered as our mother," says Mohammad Abdul Matin, general secretary of a Dhaka-based environmental group, in response to Bangladesh granting its rivers the same legal status as humans. "The river is now considered by law, by code, a living entity, so you'll have to face the consequences by law if you do anything that kills the river," declared Matin.[8]

In North America, in 2019, the Yurok Tribe in Oregon conferred personhood to the Klamath River in northern California. And in 2021, the Innu Council of Ekuanitshit and the Minganie Regional County Municipality declared the 120-mile-long Magpie River in Quebec, Canada, to be a legal person with rights. "I see the river and trees as ancestors," says a member of the committee charged with preserving the river's legal rights. "They've been here long before we have and deserve the right to live," he concluded.[9]

Across the United States and Canada, Europe, and South America, the list of similar cases continues to grow daily, in ways that Stone likely could not have imagined in 1974 when he wrote his landmark piece. The growing number of examples often draw from the Universal Declaration on the Rights of Rivers, drafted by the Earth Law Center, and endorsed by more than one hundred organizations across twenty countries, stating that rivers are living entities, entitled to fundamental rights and legal guardians.[10] "If humans are to prosper," says the center's executive director Grant Wilson, "we must transition from the old and outdated system of treating rivers as our property to a new system in which we give rivers and other ecosystems a voice within our legal system, government, and other institutions."[11]

To provide such a voice, conferring personhood upon rivers seems to be the way forward for many. The question to ask ourselves is, Why personhood? And what does that mean when it comes to the nonhuman environment? Is the practice merely a legal strategy, or is there a deeper, moral, and ecological need that deserves to be acknowledged?

Why Personhood?

"The status quo is no longer acceptable," says Yenna Vega Cárdenas, a lawyer in Quebec. "We need to have a strong water law framework, and I think legal personhood is the future for water around the world."[12]

As we have seen, Cárdenas is hardly alone in holding this view.

We need to remember that already, we have become accustomed to assigning rights to nonhuman entities for some time. Corporations, trusts, and joint ventures hold legal rights of personhood. Guardianships and powers of attorney allow for others to represent the interests of those who are unable to do so themselves. On a purely utilitarian level, appointing guardians for rivers as legal persons allows for a greater level of protection and representation of our waterways. In the words of one journalist: "Personhood raises the profile of natural landmarks by drawing attention to their beauty and cultural significance. In doing so, it builds a strong case for fostering a local economy aligned with conservationist values."[13]

The notion of personhood has a long history in Western thought. On the one hand, it can be said that it originated in ancient Rome, where the Latin *persona* meant "mask," indicating that, like a legal entity, it revealed something (though not everything) about the identity of an individual. On the other hand, a very different notion of personhood also developed through Christian thought. Building on the metaphysical roots of Aristotle and St. Thomas Aquinas, philosophers like Jacques Maritain distinguished between the notion of individual and person. As an individual, we are material parts of a whole—members of a family, of society, of a global community. But as persons, we are spiritual wholes—unique souls, endowed with intellect and eternal life.[14] More than being simply material objects, we are each sacred subjects, deserving of respect as irreplaceable wholes.

Moreover, according to the Orthodox Christian theorist Christos Yannaras, "Love is the supreme road to knowledge of the person because it is an acceptance of the other person as a whole. It does not project onto the other person individual preferences, demands or desires, but accepts him as he is, in the fulness of his personal uniqueness."[15]

Extending such a vision of personhood through a Christian lens to rivers is complicated by the fact that, typically, such interpretations of personhood were assigned solely to the soul of human beings. On such a reading, human persons could *represent* rivers in a legal court, but rivers themselves, being nonhuman, could not be afforded personhood in and of themselves.

And yet it is rivers that have been granted legal personhood in many of the examples here. The question to ask ourselves is, then, whether there is moral, ecological, and even ontological legitimacy in assigning personhood to a river beyond solely maneuvering legally for conservation purposes.

One of the most interesting insights from the past decades in my own field of environmental ethics is that the line between the human and the nonhuman has come under serious challenge. Thinkers like Tom Regan show how animals are subjects-of-a-life rather than simply objects for human use. Peter Singer reminds us that animals are sentient beings, feeling pleasure and pain, and thereby animals deserve to be morally valued. Some refer to this phenomenon of assigning moral considerability to nonhuman animals as "anthropocentric extensionism"—meaning that ethically significant human characteristics are found in animals as well.[16]

But are those characteristics found in rivers? And frankly, are we asking the right question in trying to "extend" human elements of moral considerability to nonhumans in this manner? Let's face it: on many levels, it does sound unusual to those whose hearing has been shaped by Western concepts of the self, to describe a river as a person, particularly if we understand personhood on an everyday basis in terms of autonomy, self-determination, rational minds, privacy, and so on. Might there be another way forward that summons a more fundamental paradigm shift in our way of thinking?

It's important to realize that it is not only Westerners who have a voice when it comes to defining the notion of a person. For instance, consider how in African or Bantu culture, the translation of anything that comes close to the idea of personhood immediately emphasizes one's place in a social context. In some traditions, when a child is born, the child is given one name for the father's side of the family and another for the mother's side. In this way, villagers immediately know to which groups the child belongs, simply by virtue of a name.

Other Indigenous cultures similarly situate their individuality immediately within kinship relations. I have edited books where Indigenous colleagues begin their academic contributions first by describing themselves in terms of their homes, their mother's and father's heritage, and additionally their current role in a specific university community. In such instances, one's own personhood is never defined apart from visible and invisible relationships with others.

In Japanese culture, precedence is, once again, always given to social context. "Among university graduates," explains the anthropologist Chie Nakane, "what matters most, and functions the strongest socially, is not whether a man holds or does not hold a PhD but rather, from which university he graduated."[17] One's *relations* in the world are more significant than how we are defined as discrete individuals. Moreover, there is an implicit moral invitation to look outward and be open and empathetic toward others around us. Empathy (*ohoiyari*) is held to be one of the highest virtues and signals the moral primacy of reciprocity and relation, not simply as a rational exercise but as an embodied, emotive, lived experience of being human.

Without presuming to blindly appropriate cultural traditions, one might yet ask oneself: Which lessons can I take away from these non-Western perspectives? In sitting by a river, how might my own personhood be seen to empathetically intertwine with the river as person—as living subject rather than as object separate from me?

In his novel *Colas Breugnon*, Romain Rolland describes how a main character sits quietly by the junction of two rivers. "The water flowed by calm and still with scarcely a ripple on its smooth surface; my sight was drowned in it," he writes, "as a fish is held by the hook; the whole sky was entangled by the river as if in a net, where it seemed to float with its rosy clouds caught among the reeds and grasses, and the golden sun rays trailing in the water."[18] How brilliant a description, showing how our own embodiment overflows the river, and the river itself captures the clouds, the sun, the sky! More than simply an object, the river speaks to the author, inviting an elemental sense of belonging among diverse living entities.

On a lighter note, recall the childhood treasure *The Wind in the Willows*, in which the Mole asks the River Rat: "And you really live by the river? What a jolly life!" Whereupon the Rat replies: "By it and with it and on it and in it.... It's brother and sister to me, and aunts, and company, and food and drink, and (naturally) washing. It's my world, and I don't want any other. What it hasn't got is not worth having, and what it doesn't know is not worth knowing." He concludes, "Lord, the times we've had together!"[19]

Once again, the passage provokes our thinking about being with the river in more than a utilitarian, legal sense of personhood (useful though that can be) and other than as an instrument or resource for human use. Instead, we are reminded of our watery origins, of our deep, elemental belonging to water. The grace of a river, as ecosystem, as watershed, reminds us that we belong to something larger, something that we did not create but within which we must find our proper place.

If we are going to meaningfully speak about rivers as persons, it will be to recognize that we ourselves are bodies of water, which means, to quote the hydrofeminist Astrida Neimanis, "refusing a separation between nature and culture, between an environment 'out there' and a human subject 'in here.'" It means "learning from water. Water is a connector." [20]

To return to our opening story, let us remember that each child emerges from the fluidity of the womb, nourished, surrounded, and indeed elementally constituted by water.

So—in the end—are rivers persons? Legally, in many cases now, yes. There is some utility in assigning guardians who can oversee conservation efforts. Even corporations are beginning to see that their business models need to be reconfigured in response to environmental pressures: witness the retailer Patagonia's recent announcement that Planet Earth "is now our only shareholder." [21]

Does that mean that rivers are persons in the typical Christian, Western sense of the word, where we might anthropomorphize them, and awkwardly assign human characteristics? My sense is no.

But perhaps there is yet another way forward—a way to rethink the ecological personhood of rivers as living subjects, overflowing our own personhood, constituting an invitation to rethink our own place in this watery world, and find a better balance in relation to it. Perhaps rivers can remind us that we are never alone but belong always, essentially—indeed, elementally—to the watery planet that sustains us all.

notes

1. See Bronagh Kieran, "The Legal Personality of Rivers," *EMA Human Rights Blog*, http://www.emahumanrights.org/2019/01/16/the-legal-personality-of-rivers/.
2. See Christopher Stone, *Should Trees Having Standing? Law, Morality and the Environment*, 3rd ed. (New York: Oxford University Press. 2010), 3. For an excellent, recent survey of the current state of environmental jurisprudence, see David Boyd, *The Rights of Nature: A Legal Revolution that Could Save the World* (Toronto: ECW Press, 2017).
3. Grant Wilson and Darlene May Lee, "Rights of Rivers Enter the Mainstream," *Ecological Citizen* 2, no. 2 (2019), 184.
4. Cited in Ashley Westerman, "Should Rivers Have Same Legal Rights as Humans? A

Growing Number of Voices Say Yes," *NPR*, August 3, 2019, https://www.npr.org/2019/08/03/740604142/should-rivers-have-same-legal-rights-as-humans-a-growing-number-of-voices-say-ye.

5. For more, see the Whanganui Iwi (Whanganui River) Deed of Settlement Summary | New Zealand Government (www.govt.nz).

6. See Brian Roewe, "In Colombia's Chocó Region, 'River Guardians' Protect the Rights of the Rio Atrato," *EarthBeat*, January 28, 2022, https://www.ncronline.org/news/earthbeat/colombias-choc-region-river-guardians-protect-rights-rio-atrato; Community Environmental Legal Defense Fund, "Colombia Constitutional Court Finds Atrato River Possesses Rights," May 4, 2017, https://celdf.org/2017/05/press-release-colombia-constitutional-court-finds-atrato-river-possesses-rights/; Anima Mundi Law Initiative, "Rights of Nature Case Study: Atrato River," http://files.harmonywithnatureun.org/uploads/upload1132.pdf#:~:text=In%20November%202016%2C%20the%20Constitutional%20Court%20of%20Colombia,Afro-descendent%20groups%20for%20the%20protection%20of%20constitutional%20rights.

7. For more, see Zoe Cormier, "In Ecuador, Rivers, Plants, and Animals Have Rights," *Globe and Mail*, October 11, 2008.

8. For more, see Westerman, "Should Rivers Have Same Legal Rights as Humans?"

9. Chlor Berge, "This Canadian River Is Now Legally a Person. It's Not the Only One," *National Geographic*, April 15, 2023, https://www.nationalgeographic.com/travel/article/these-rivers-are-now-considered-people-what-does-that-mean-for-travelers.

10. For more, see IUCN World Conservation Congress, "The Universal Declaration of the Rights of Rivers: Updates on a Global Movement," https://www.iucncongress2020.org/programme/official-programme/session-52640#:~:text=The%20Universal%20Declaration%20of%20the%20Rights%20of%20Rivers,wishing%20to%20join%20the%20rights%20of%20rivers%20movement.

11. See Pier-Olivier, Monti Aguirre, and Grant Wilson, "Why Recognize a River's Rights? Behind the Scenes of the Magpie River Case in Canada," International Rivers, March 15, 2021, https://www.internationalrivers.org/news/why-recognize-a-rivers-rights-behind-the-scenes-of-the-magpie-river-case-in-canada/.

12. Quoted in Ben Andrews, "A Global Movement Is Granting Rivers Legal Personhood. Could the Gatineau River Be Next?" *CBC News*, April 2, 2023, https://www.cbc.ca/news/canada/ottawa/gatineau-river-legal-person-1.6794975.

13. Berge, "This Canadian River."

14. For a concise summary of Maritain's views, see Joseph W. Evans, "Maritain's Personalism," *Review of Politics* 14, no. 2 (April 1952): 166–77.

15. Christos Yannaras, "Person and Individual," in *The Freedom of Morality* (New York: St. Vladimir's Seminary Press, 1984), 23.

16. For a nice introduction to the field, as well as excellent summaries of Regan's and Singer's works, see Joseph R. DesJardins, *Environmental Ethics: An Introduction to Environmental Philosophy* (1993; Belmont, CA: Wadsworth Publishing, 2012).

17. C. Nakane, *Japanese Society* (Harmondsworth, UK: Penguin Books, 1974), 3.

18. Romain Rolland, *Colas Breugnon Burgundian* (New York: Henry Holt and Co., 1919), 99–100.

19. Kenneth Grahame, *The Wind in the Willows* (independently published in Bolton, Ontario: Amazon, n.d.), 6-7.

20. See the interview with Neimanis, by Richard Bright, in *Interalia Magazine* (September 2018), https://www.interaliamag.org/interviews/astrida-neimanis/.

21. Suzanne Bearne, "The Firms Giving Nature a Stake in Their Business," *BBC News*, July 13, 2023, https://www.bbc.co.uk/news/business-66132669.

The Empathy of Rivers:
Dreaming with the Río San Pedro

Lisa María Madera

I have been dreaming about plastic. I am at an overnight retreat or conference, standing alone in an open kitchen, talking to newfound friends. I feel a hair tickling the back of my throat and excuse myself, retreating to the kitchen to find a glass of water. The water does not resolve my discomfort. I reach into my throat and grab what I think is the hair and begin pulling. It is not a hair at all but a plastic fishing line. I tug on it and realize that it is hooked on to something down my throat. I bend over and hide beneath the counter so that my new friends won't see that I am caught in an unflattering predicament. I tug and tug. Soon I am pulling out plastic bags attached to the fishing line. There is a blue- and white-striped bag. A green one. A pink one. On each bag, I can read the labels and logos. I pull and pull and pull, tugging out plastic after plastic after plastic.

It's noon in Quito. Thunder crackles overhead. The winter rains have come. The sky rivers from the Amazon drench the highland plains. A cold Andean wind blows in from Cotopaxi. Troubled lately, the sleeping giant is active again. The massive volcano has been breathing out ash and gas.

Here, on the flanks of Pichincha a thick fog wraps the city. I have been on the phone to my friend Marilyn in Northern California. We are meeting for coworking, together creating a rhythm of work to be able to turn our grief and concern, our spiraling earth anxiety, into productive action. We are working on materializing our dreams. She asks me how I am. "Fantastic! I'm doing great!" I say.

My exuberance overwhelms me, and I pause, embarrassed, "How are you?"

"Not so good," she says.

"It's the moon," I offer. The full moon always unsettles me.

"No," Marilyn says, "the weather."

And I remember. It's raining in California. I have just read this morning in the news that a cyclone bomb is blasting California coastlines. *A cyclone bomb?* I have never heard of such a thing. Delivered by the Pineapple Express, the Hawaii sky river, this cyclone bomb is yet another new and terrifying weather phenomenon to add to our growing climate-change lexicon. Whole neighborhoods were evacuated in the worst storm California has experienced in the past few years.

"The winds were terrifying," Marilyn murmurs.

Marilyn lives in the flats, close to the bay, and the nearby creek is filling her neighbor's basement. Ironically, the heavy rains won't resolve California's water crisis. Attuned to the well-being of Earth's rivers, I have been following the news that the Colorado River is about to run dry.

"It's here!" Marilyn says to me, "The climate catastrophe is here."

It feels like an official announcement, Marilyn's morning proclamation. As if with her words, we stepped over the edge.

I Want to Tell You a Love Story

I want to tell you a love story about a girl, woven to womanhood through the compelling voices of water, the wondrous *wak'a* rivers born from *wak'a* mountains pouring toward the sacred sea.[1]

The story I want to tell you is a river itself, with many streams that feed it. Part of this story I will sketch in shorthand and hope to return at another time and place to fill in details for you. But for

now, the sketched-out map of this story will lead us to the trans-
formative banks of the San Pedro River on the edges of the Quito
inter-Andean valley, where we will pause for a moment to wonder,
learn, and reflect.

I was born in Quito, in the middle of the world on the flanks
of the volcano Pichincha. My father, a Minnesota Swede, was
an Evangelical medical missionary serving in hospitals in the
Ecuadorian Andes and Amazon. From the ages of two to five, and
from eleven to eighteen, I lived in Quito. From six to ten, I lived
in the Ecuadorian Amazon, in the rainforest, in a military town
named for Shell Oil. Interwoven in months and a few years in be-
tween, we would go back to visit my father's family in Minneapolis.

My beautiful, sharp-witted mother, an educator trained in
one-room schoolhouses in the ranching wilds of Nebraska, gave
me my first book when I was two or three. Through it, my nose
quickened to the talcy smell of white flowers perforated into white
pages. My hand learned to trace the stubble of my father's chin and
reach for the silken fur of rabbits as I read through the textures of
the page. That book was my second interactive sensory storybook.
My first? Nature. Pacha Mama, herself.

I was a child raised with my knees pressed to the flagstone
tracing out the phosphorescent trails of snails across hand-cut
Andean stone. Trained in a land laced by fragrance, my nose quiv-
ered to *colada morada*, Andean lupines, the liquid sweetness of hot
chocolate, and the mesmerizing, hypnotic wonder of eucalyptus
salting the wind. The first river I remember meeting as a child
in Quito was the San Pedro, although at the time I didn't know
their name. I do remember how the river flowed past the pools at
Cununyacu, where we would go occasionally with my family on
picnics to swim.

In Shell-Mera, I learned to think in the cold currents of
Amazonian rivers—the Alpayacu, the Pindo—clambering across
river stones. I used to scramble over great boulders barefoot, my
toes searching out unexpected niches to climb up onto the top of

the looming rock. I have lain on immense, hot white boulders after swimming in cold waters, and I have thrown out my arms like a star or an angel and sighed with the deepest delight. My back to the boulder. My face with the breeze tinkling my cheeks. Oh, I used to stop to feel the wind on my knees! Deeply ingrained, the sharp, cold scent of river sand still causes me to pause, quickens my heart, and stirs up a whirlwind in my belly of longing and loss and joy and delight and fright and liquid expectation all in one.

My father taught us to pay attention to water. As a family doctor who had seen his share of dangerously dehydrated children, my father was understandably fastidious about the water we drank or used to brush our teeth. Odd, isn't it? Except for piranhas or the dangers of drowning, I don't ever remember him being worried about us swimming in the Amazonian rivers. Not like now, when oil, gold mines, pig farms, textiles, palm plantations, factories, and plastics have intoxicated the water and suffocated the rivers. We have learned to pause before stepping into a river. You worry about scalding your skin in heavy metals or agrotoxins or thick, black petroleum crude.

Lately I have been worrying about our rivers. In the fifty-one years since oil has come to Ecuador, many of Ecuador's rivers have been poisoned. The San Pedro and the Pindo, for example, are now both unswimmable.

I've been listening to water since before my conception, but only recently have I begun to understand myself as an ally or a kind of relative to rivers, a daughter, a sister, a niece, maybe, or a cousin. There is a Kichwa description for this in the Amazon, *yacu warmi*, water woman. Only recently have I begun to hear the river calling out my name.

We Are All Rivers

My Guayllabamba Waterkeeper friend Patricio Chambers tells me: "We are all rivers. Our lives carry us out to our final moments when we will pour out our hearts to the sea."

If you understand a story line and if you understand rivers, you will understand its flows and intentions. You will recognize its white water, black water, turbid water, and whirlpools. A story line is a landscape that has its own peculiar topography, just as our lives do. Our lives, too, mirror the sensuous curves of the mountains, the roaring curve of the rivers. Our lives, too, are full of currents that overtake and overwhelm us and sometimes carry us along like Huck Finn with his back on that raft trying to find a way to escape the narrative force of his own history, the history of the Mississippi River and the history of those around him.

The landscape of our lives yields its own rise and fall of currents. We are shaped by stories filled with information, just as the wind and the currents of rivers and oceans carry the choreographed music of life, information so rich and subtle that our only response as our bodies step into those currents and feel the tickling rippling voice of the water is AH! AH! AH! These were my daughter's first words, as a baby strapped to our backs, when she pointed for us to take her down to the shores of the Río San Pedro, the first river she ever met, "Ah! Ah! Ah!"

The delicious delight of water encountered in a healthy stream, a mineral healing spring, a volcanic hot spring, a healthy river, a lake, or an ocean is intoxicating. Pause for a moment and let your body remember. Can you feel it? The water rippling across your body? The water soothing your skin? The rhythm of the water surrounds you. The rush or the stillness of the currents lull you while water fingers twist and twirl your flowing hair.

It is no wonder that mythical stories of water conjure water lovers that whisper to us in our ear. The voices of rivers and oceans are magical and miraculous. If you have ever fallen asleep to the sound of the sea, you will remember how the ocean's voice becomes part of your heartbeat. You will remember the feel of the sound of the surf rushing through the rivers of your body, the sound of the ocean beating through the chambers of your aching heart. Your body remembers tuning itself to the water. Your body remembers letting go of its sorrow and its heartache to the tumult and roar of the sea.

Rivers run through us. They do! And they weave us together. The rivers that flow through the contours of our body carry water and blood, healing platelets that surf through our body and our arteries, water flowing from tongue tip to pee point; water coming in and water going out of the body of our land and the land of our body. We are always in the process of circulating nourishment and toxins, but now that we have stopped listening to the voice of the rivers, we have thoughtlessly changed the course of rivers through concrete, dams, and diversions. We turn our backs to our rivers while we overwhelm them with plastics and poisons. We don't even remember that the rivers on the outside of our bodies are intimately interconnected with the rivers that course beneath our skin.

There is no separation between us. We are an infinite cycle of circulating systems of co-creating, co-sentient beings woven together through a pulsing collective in which knowledge and wonder travel and transform between us. This spiraling cycle of music and matter, this *wak'a* wondrous world, carries within it the first droplets of light and the very first sounds that harmonized time. How this happened—physicists, evolutionary biologists, poets, philosophers, historians, and theologians, along with many questioning four- and five-year-olds, are all doing their best to find out.

We are, in fact, a walking, swimming, crawling, creeping, flying, oozing, sliming water collective. Just like our watery planet, most of our bodies are made up of H_2O. Brought into being within the watery world of our mother's womb, baby humans are born water-rich, with water filling 78 percent of our bodies. It isn't easy to adjust to the dry outer world. As we come into the dry world of air and begin to adjust to that new state of being, water currents flow so forcefully through our new bodies that we cannot contain them. As babies we cry and leak all the time. It takes time to learn to control the flow of water pouring through us.[2]

Rivers run through us. They do.

Our human-river lives are inextricably intertwined.

At the same time, rivers exist with their own individual integrity, as unique interlocking entities. In Ecuador, we understand this, and that is why our rivers have rights.

Ecuador, the Rights of Nature, and *Sumak Kawsay*

Fifteen years ago on October 20, 2008 in the charming coastal town of Montecristi, famous for making the beautiful and misnamed Panama straw hats, Ecuador's congress gathered and passed a new constitution, encompassing a multicultural, plurinational state reflecting Andean and Amazonian cosmological connections to Earth.

As the first in the world to establish and protect the inherent rights of Nature, Ecuador's landmark constitution establishes the sovereignty of Indigenous Peoples as independent nations, it articulates the rights of vulnerable peoples, and reiterates citizens' rights to *sumak kawsay*, or *buen vivir*, an Andean vision of well-being that promotes a communal democracy and recognizes the interrelated relationships between Pacha Mama and all living beings. Article 14 of Ecuador's 2008 constitution recognizes the "right of the people to live in a healthy and ecologically balanced environment that guaranties sustainability and a good life, Sumak Kawsay." To achieve *sumak kawsay*, the constitution highlights citizens' rights to information, participation, and justice.

Notably, our constitution names and acknowledges Pacha Mama, Mother Earth, as a living, evolving being. Article 71 states that "Nature or Pacha Mama, where life is reproduced and realized, must be respected for its inherent right to existence including the maintenance and regeneration of its life cycles, structures, functions, and evolving processes."[3]

Article 72 goes on to describe Pacha Mama's right to renewal, in the event that Nature's integrity has been violated or compromised: "Nature has the right to restoration." The constitution holds the state responsible for "establish[ing] the most effective mechanisms to achieve restoration and adopt suitable measures to eliminate or mitigate harmful environmental consequences."[4]

It takes a while for a constitution to take root in the land, and after fifteen years and scores of fascinating and notable landmark cases, Ecuador leads the world in bold legal actions defending Nature's rights, especially the rights of rivers.[5]

Initially, cases moved very slowly before the Supreme Court would decide on behalf of the river's rights. This is the case of the Río Dulcepamba (2007 to 2020) whose currents and natural course have been violently altered and confused by the San José del Tambo hydroelectric project.[6] Recently, however, legal currents for the rights of Nature seem to be gaining momentum, and there are cases in which justice is being delivered more swiftly, as in the case of the Río Monjas. On January 19, 2022, after two years of process, Ecuador's Supreme Court determined that the Río Monjas "is sick and has lost its ecologic equilibrium and requires restoration," determining that the river's rights had been violated by the Municipal District of Quito (DMQ), obliging Quito to develop a "green blue" city ordinance providing for the care and healthy management of water and natural ecosystems within the city. By July 4, 2023, eighteen months after the ruling, the DMQ approved the Ordenanza Verde Azul.[7]

In Ecuador, we are witnessing rising citizen support for the rights of Nature. In the fortuitous time between the blue moons of August 2023, something extraordinary happened. After an intense and emotional campaign, on Sunday, August 20, 2023, Ecuador made planetary history.

Mark your calendar! Write it down!

On August 20, 2023, Ecuador voted in landslide decisions in two separate public referenda (one municipal and one national) to defend the rights of Nature in two megadiverse national parks and UNESCO biosphere reserves. Recognizing the importance of Quito's water source, 68 percent of Quiteños voted to protect their rivers and keep mining out of the Chocó Andino. And then, nearly

59 percent of Ecuadorians voted to keep oil in the ground in the Amazonian reserve of Yasuní National Park, among the most biodiverse corners of our planet. Yasuní means "holy land" in Huaorani, and the site contains the ancestral territories of the uncontacted Tagaeri and Taromenane, people who live in voluntary isolation in the forest. There are many reasons this life-giving place is sacred, among them that Yasuní has more species of trees per hectare than any place on Earth.[8] This year on October 20, 2023, the fifteenth anniversary of the declaration of the rights of Nature in Ecuador's constitution, we have much to celebrate.

These celebratory currents lead us back to the banks of the Río San Pedro, to our river, where over the past two years, throughout the course of fourteen *mingas*, I have joined an amazing, inspiring, intergenerational, intercultural group of friends and allies who together make up the Colectivo Rescate Río San Pedro. Together we organize *mingas*, communal eco-educational, art-filled, interdisciplinary river-rescue events at which we work to heal our river. In our *mingas* we educate and celebrate and defend our river's integrity and inherent right to live free of textiles, plastic, agrotoxins, sewage, industrial waste, and all the piles of human trash that asphyxiate its currents.

The *minga* (from the Kichwa word *minka*) is an Andean ancestral social and educational technology that weaves community while caring for the earth. The Inca, who made *mingas* famous, used them to construct extensive roads and irrigation canals. Similar to how Amish neighbors might come together for a traditional barn raising, *mingas* harness and channel individual talents in collaborative collective work to accomplish communal goals.

In our collective, we work together with a network of allies that include Guayllabamba Waterkeeper, Arcandina (an innovative, award-winning, green puppet show), Rotary Club Distrito 4400, local schools and universities, ecumenical communities of faith, businesses, and members of local, provincial, and national governmental agencies. We come together to care for our river while providing an educational, art-filled, festive, and celebratory community event. At the end of our *mingas*, we share a *pamba mesa*, a meal in which we all bring beautiful healthy food to share. Our collective has become famous in Quito for our educational and environmental impact. In the past two years, we have harnessed more than 8,500 volunteer hours to remove over twelve tons of garbage from the river. Last year on World Water Day, we expanded our *minga* to seven provinces with Ecuador's Ministry of the Environment and to eight countries with the Waterkeeper Alliance. This next year, with Rotary Distrito 4400, Waterkeeper Alliance and The Nature Conservancy, our *minga* will go worldwide, celebrating World Water Week in March 2024 with the Minga Mundial.

Our *mingas* create a communal healing space that circulates and transfers knowledge, technology, and best practices. It is a focused, productive, interdisciplinary, social, festival space in which community ties are woven through setting objectives and meeting them through the strategic allocation of resources, including human talent, experience, and knowledge. It is a family space and an intergenerational, intercultural community science space that celebrates connection to one another and the Earth. It is a space where we interact with the river, where we listen and learn from the river as we work to care for the Earth. The *minga* is a co-creative space in which we co-create with one another and with the Earth. It is a festival, a space that creates physical and spiritual exchange. It is a space marked by rituals and blessings for the river and the land. It is a space that weaves alliances, a ceremonial space in which we seal agreements. It is a networking space in which we meet really interesting new friends. It is a space filled with art and music and beauty. A joyful space where we connect and care for one another, the river, the land, and our interconnected web of relations, both human and more than human. It is a place that invites communities of faith, artists, activists, scientists, collectives, the academy, eco-entrepreneurs, nongovernmental organizations, and government officials to come together and co-create and co-design projects that bring about *sumak kawsay, el buen vivir*. In a very real way, the *minga* functions as an ancestral social technology materializing *sumak kawsay*. When we celebrate the *minga*, we live well.

We have a saying in our *colectivo*. *El río nos llamó. El río nos unió*. The river called us. The river united us. And it is true. When you work for a morning with your hands in the river with a group of strangers to figure out together how to get a car tire off of an opposite bank without getting wet because the river is poisoned and toxic, you learn mighty quickly how to make friends.

Our mission is clear, we want to swim in our river again.

"Somos locos. Somos unos locos soñadores, soñadores con nuestras botas plantadas firmemente en el río," my friend Pablo Palacios, the award-winning documentary filmmaker and cofounder of Arcandina says to me. "We're nuts. We are a bunch of crazy dreamers, dreamers with our boots firmly planted in the river."

The climate crisis is a water crisis. And in that knowledge lies our pathway forward. In our August elections in Ecuador, we voted for the lungs of the world. We voted for the birds and the bees and the bats of the world, for the creatures of the night to have the right to darkness. We voted for the waters of the world. For the people of the water, to have their right to the healthy sea. We voted on behalf of our rivers, on behalf of our Río San Pedro.

The climate crisis is a water crisis, and as such, it is a *wak'a* crisis. To find balance again, we must understand that rivers run through us and that together, we are part of a sacred collective current of life-giving co-creative exchange. This is the way to come into relation with the wondrous *wak'a* world. We must learn to listen again to the river, learn to listen and respond when the rivers call our name.

acknowledgments

I am grateful to Gavin Van Horn for the unexpected opportunity to think and write about water. The things I learned this year have transformed me. Thank you Ingrid Stefanovic for your timely, transformational edits. I am grateful to Marilyn Paul for her steady companionship navigating the currents of climate change. Thank you Harry Greene for your writing wisdom. Deepest gratitude to the Colectivo Rescate Río San Pedro, especially my colleagues on the *directiva*, Maribel Pasquel, Patricio Chambers, and Pablo Palacios. Most of all, deep gratitude to our beloved Río San Pedro for the privilege to learn along their side.

notes

1. For a description of the Andean notion of *wak'a*, see my essay "The Empathy of Birds: Lessons from Pacha Mama in the Face of Despair," *Minding Nature* 13, no. 3 (Fall 2020): 74–84, https://humansandnature.org/the-empathy-of-birds-lessons-from-pacha-mama-in-the-face-of-despair/.
2. See Water Science School, "The Water in You: Water and the Human Body | U.S. Geological Survey," US Geological Survey, May 22, 2019, https://www.usgs.gov/special-topics/water-science-school/science/water-you-water-and-human-body.
3. See the text of Ecuador's Constitution, at the website of the Center for Democratic and Environmental Rights (http://centerforenvironmentalrights.org).
4. A full version of the constitution is available at "Ecuador: 2008 Constitution in English," *Political Database of the Americas*—Georgetown University, January 31, 2011, https://pdba.georgetown.edu/Constitutions/Ecuador/english08.html.
5. For an analysis of failed and successful rights-of-Nature lawsuits in Ecuador between 2009 and 2015, see Craig Kauffman and Pam Martin, "Testing Ecuador's Rights of Nature: Why Some Lawsuits Succeed and Others Fail" (presentation at the International Studies Association Annual Convention, Atlanta, March 18, 2016), http://files.harmonywithnatureun.org/uploads/upload471.pdf.
6. For a history of landmark decisions, see the website of the Observatorio Jurídico de Derechos de la Naturaleza (https://www.derechosdelanaturaleza.org.ec).
7. For a review of the Río Monjas case, see "El Río Monjas, que recorre la ciudad de Quito, reconocido como sujeto de derechos," Observatorio Latinoamericano de Conflictos Ambientales, March 28, 2022, https://olca.cl/articulo/nota.php?id=109331.
8. "Scientists Identify Ecuador's Yasuni National Park as One of Most Biodiverse Places on Earth," *ScienceDaily*, January 19, 2010, https://www.sciencedaily.com/releases/2010/01/100119133510.htm.

Water, Gender, Justice: Women's Anti-Hydropower Activism in Turkey

Özge Yaka

P eter Münch begins his *Süddeutsche Zeitung* article on a group of Palestinians who traveled to the Mediterranean coast, which is geographically so close—yet so far away—this way:

> And suddenly Fatmi stands with her feet in the sea...
> Wherever she looks, everything is blue, everything is the
> sea. She can feel it, she can hear it, she can smell it. "You are
> speechless when you see this beauty," she says. "I love the
> waves, the sound of the sea, the colors. I love it all." Fatmi
> wanted to go to the sea. Absolutely, and for a very long time.
> Now she is there, together with her husband and her son, and
> it is the first time for all of them.[1]

The sea is nearby, only eighty kilometers away from the village where Fatmi lived all her life; but it is far away, located within the borders of Israel.

As a Palestinian living in the West Bank, it is not easy for Fatmi to step on Israeli soil. That is why she is standing by the sea for the first time in her life, even though she had wanted to do so for as long as she remembers. The long-awaited encounter overwhelms her at first sight, as she is immersed in a multisensory experience of the Mediterranean Sea.

Not all of us risk a dangerous journey like Fatmi to have a firsthand experience of water bodies, but most of us are drawn to oceans, seas, lakes, ponds, rivers, and creeks. We often spend our holidays on the coast or by a lake or river. We visit spas and thermal springs to relax and rejuvenate. We go to great lengths for a home with a sea, lake, or river view. Immersing one's body in water, watching the movements of water bodies, hearing the waves or a river's flow, drinking from a spring—firsthand sensory experiences of water are highly affective and desirable, even though they are all-too-rarely discussed in a scholarly manner.

Our relationship with water certainly has an instrumental dimension: it is life giving and life sustaining, not only for humans but for all living beings. It is, then, understandable that the literature on water is dominated, overwhelmingly, by discussions of water as a vital *resource*, a term that focuses on the aspect of utility in the fields of water management, governance, access, development, security, and sustainability.

Utilitarian framings of water, however, render invisible the deep, living connection between water and our bodies and our utter dependency on water, both fresh and salinated, in every facet of our lives and our well-being. The story I tell in this essay—the story of the women of Turkey's East Black Sea region who resist the small-scale hydroelectric power plants (HEPPs) that invade their villages and valleys and "steal" their streams—is about the strength of such sensory, affective, and ultimately social connections to water as an elemental relationship, which is mostly ignored by the scholarly literature.[2]

Standard arguments about local community struggles against various forms of resource extraction and dispossession (such as mining, oil drilling, fracking, and hydropower) share the tendency to frame water as a utility on which lives and livelihoods depend. The struggles, it follows, are motivated by this immediate biological and economic dependence or, in the cases of Indigenous struggles, by belief systems in which water bodies have a central role.

I certainly do not deny the importance of livelihoods and belief systems for many local, especially rural and Indigenous, communities fighting to protect water bodies around the world. However, here I introduce an alternative framework, drawing on an empirical case of women's activism against HEPPs in Turkey's East Black Sea region. I argue that fundamental, embodied relations to waterways are central to such activism and much more meaningful than simple utilitarian reasoning allows.

Gendered Activism against HEPPs in Turkey

Run-of-the-river HEPPs are substantially different from hydroelectric dams, as they require little or no water storage. Instead, they capture the kinetic energy of the natural flow of water as it runs down steep slopes. Typically, water from upstream is diverted to electricity-generating turbines by a weir or a pipeline and then is released back into the river's downstream flow. For this reason, HEPPs are often presented as an eco-friendly alternative to hydro dams and a less ecologically harmful way of producing renewable energy, despite the emerging evidence of their realistic ecological impacts.[3]

During the 2000s, under the Justice and Development Party (Adalet ve Kalkınma Partisi, or AKP by its Turkish initials), Turkey decided to make extensive use of small-scale, run-of-the-river HEPPs. In accordance with AKP's construction- and energy-based growth strategy, sometimes referred to as "bulldozer capitalism," private construction and energy companies have been given "extraordinary latitude to evict villagers, expropriate private land, clear state forests and steamroll normal planning restrictions to meet the target of four thousand hydroelectric schemes by 2023."[4]

HEPPs started to pop up in the remote villages and valleys of the country, especially in the East Black Sea and Mediterranean regions, and in the East and Southeast Anatolia to a lesser extent, where the small and middle-sized rivers have naturally steep slopes. By 2007–2008, it became clear that HEPPs were destroy-ing both river ecosystems and natural habitats especially when they are built extensively, without adequate impact assessment, regulation, or monitoring. In Turkey, dispossessed riverside com-munities were left with dry streambeds when river waters were diverted through pipelines for kilometers.[5] Riverside communities started to organize themselves locally against HEPPs, and local movements emerged in many different parts of the country, slowly leading to the formation of regional and national networks of com-mitted activists.

Even though the anti-HEPP movement is not a women's movement per se, women shape the specific character and public image of the movement with their bold, committed, and radical activism.[6] Especially in the rural East Black Sea region, where re-sistance to HEPPs is concentrated, women have been highly visi-ble, protesting in their traditional clothing of *şalvar*, long skirt, and headscarf, keeping guard at the construction sites and blocking construction equipment—even physically confronting the military and the police when necessary.[7] Contrary to men, who assume a more moderate language, women adopt a radical position, talking about killing and dying to protect the rivers, which is not unusual in local environmental struggles.[8]

Take, for example, Selime, a middle-aged woman whose words I recorded during our conversation on her terrace in the village of Arılı, in the East Black Sea region: "We, as women, won't allow them to construct a hydropower plant here. We don't even count on men.... Bring them on, if they dare, if any brave fellows think they can come here... let them try. We will cut them to pieces. We know how to use guns as well. We take the risk. They really shouldn't force us. Don't make people go mad."

Selime's courage and commitment are similarly shared by many other women with whom I spoke during my field research—and also by many others who have been interviewed by different journalists.[9] No one is motivated by an immediate economic dependence on river waters, as scholars of feminist environmentalism and political ecology would assume. In fact, livelihoods in the East Black Sea region do not depend on river waters, because they are not used for agriculture or daily household needs: rainfall alone sustains monocultural tea cultivation (and hazelnut crops in the western parts of the region). In other words, rivers are not "natural resources" for East Black Sea women, and their struggle cannot be captured by terms like *resource conflicts*.

Instead, what drives their activism are the intimate corporeal connections that they have with river waters and the bodily memories of such connections that are anchored in place.

When I asked East Black Sea women why they oppose HEPPs, they talked neither about livelihoods nor macropolitical struggles, technical data, or scientific information, as men tend to do.[10] Instead, they told me about the river itself and their everyday bodily interactions with it—about how they grew up by the river, how they are used to sleeping with and waking up to the sound of it, how they love immersing their bodies in its waters, and how the sight of its flowing, cascading waters defines the place, whether a village or valley.

"We live here, in this narrow valley, only with the joy of the river [*derenin neşesi*]. When it is gone, it means we should also go," said Refiye, a villager from Yaylacılar. "We live here with the river. We look at the river every day. We cannot live without it," adds Ayşe, from Ulukent. Such testimonials are often repeated. "Rivers are our celebrations [*şenlik*]," said Ülker from Aslandere. "I went by the river today before I prayed. I just sat by it and watched it. It was so beautiful, greenish blue. I watched the fish swimming in it. I stayed there for some time and returned to pray," she said.

"The river is life and soul [*hayattır ve candır*] for us. I would probably feel utterly empty if I did not hear the sound of the river," reflected Seniye, also from the village of Aslandere.[11]

Women similarly talked about the memories of their parents by the river and the sight of their children and grandchildren playing in the same waters where they once played. "I see my mother and my father by this [Arılı] river; every time I look at the river, I remember them," said Semra, from the village of Gürsu.

"They are my grandchildren," Nuran said. "Look how happy they are in the river," said Nuran, another middle-aged woman from the village of Gürsu, pointing out three young children playing in the [Arılı] river, shouting and playing with joy during our interview.

For these women, the river is neither a mere resource that can be utilized and traded nor an abstract political, cultural, or belief-related symbol. The river is a nonhuman, environmental entity (like a neighbor, like a relative, they would say) with which they live in a close relationship that is established through constant bodily encounters and sensory-affective connections.

Children playing in the Arılı River, photo by the author

This (human-nonhuman) relationship with the river is espe-
cially strong for women because the gendered division of labor in
the region puts them in a close(r) relationship with their imme-
diate environment. Working in the tea and hazelnut fields, which
are typically located by small and middle-sized rivers that flow
from Kaçkar Mountains to the Black Sea, is deemed as women's
responsibility. Because of the proximity of fields to local rivers,
agricultural work involves everyday interaction with river waters
and enables an intimate relationship between women and river
waters. And it is that strong, close, and intimate relationship, es-
tablished through everyday practices and interactions, that moti-
vates women to oppose HEPPs in a radical and committed manner.

Women's political agency, in this case, is shaped by the inti-
mate corporeal connections they have with river waters and by
their bodily memories of such connections that are anchored in
place. In this sense, women's anti-hydropower activism in Turkey's
East Black Sea region illustrates "how agentic properties emerge
and endure within corporeal experience."[12] The case also demon-
strates that the struggles to protect water bodies against energy and
infrastructure projects are not always driven by the instrumental
use of water to sustain livelihoods; the centrality of water bodies
to our world of lived experience, to our *lifeworld*, can also inspire
such struggles.[13]

More-Than-Human Lifeworld and Socio-Ecological Justice

To describe the centrality of rivers to the lifeworld (*Lebenswelt*), of
riverside communities, especially for women, means recognizing
the more-than-human character of the places where our lives
unfold. As reflected in the German origins of Husserl's *Lebenswelt*
(lifeworld) or Uexküll's *Umwelt* (environment, surroundings),
natural entities such as waterways are constitutive of the world
of experience in which self, subjectivity, and agency are formed.
Nonhuman entities such as rivers are an essential part of the

intersubjective universe of experience, encounter, and interaction. What we call "natural" or "ecological" is, in fact, an element of the "social."[14]

This notion of eco-sociality is central to the justice claims of anti-HEPP activists in Turkey, especially in the East Black Sea region, where rivers are seen not as natural resources but as nonhuman entities with which people live.[15] Indeed, living in a riverside village in the region is living *with* a river, as people work, unwind, celebrate, socialize, sing, cry, and fall in love, always in continuous corporeal interaction with river waters. In this sense, their struggle is not for resources but for coexistence.[16] Such coexistence means living with the nonhuman, environmental entities that they value, in places that they constructed free from institutionally sustained destruction, degradation, pollution, toxification, and commodification of ecological systems, habitats, and entities.

The environmental justice (EJ) movement and the related body of EJ scholarship have famously linked environmental issues to notions of power and inequality. The two stress the unequal and unfair distribution of environmental goods and hazards—a distribution that structurally disadvantages people of color and poor populations.[17]

Even though I value the EJ movement and scholarship highly, I believe the struggles of the rural East Black Sea communities, especially the struggles of women to protect their rivers, call for an alternative notion of justice. This alternative notion, which I call socio-ecological justice, corresponds to eco-sociality and the right of coexistence and signifies the interconnection between the so-called social and ecological realms. In other words, socio-ecological justice maintains that our intrinsic and intimate relations with the nonhuman world are an essential part of our well-being and are central to our demands to pursue a fair and decent life.

Thus, what is at stake here is not merely a matter of extending the community of justice to include nonhuman environments, as the concept of ecological justice attempts to do.[18] It is also a matter of actively incorporating human-nonhuman relational ontologies and corresponding ethical practices into our understanding of justice.[19]

Muhammet Kaçar reports in the daily newspaper *Hürriyet* about a group of women who blocked the road in the Arılı Valley and who talked to the experts appointed by the court for the HEPP case.[20] One of them told him: *"Dere bizim eşimiz, biz derenin eşiyiz—* The river is our spouse; we are the river's spouse." The word *eş* in Turkish is used for "partner" or "spouse" in everyday life, but it also has the connotation of being one's equal, fellow, or companion. *Dere* (small river or stream) can be all of them—a partner, a relative, a brother, a sister, a friend, a neighbor, a relative, a companion, all in all an equal—a nonhuman person-being-entity with whom one has a lifelong, intimate, sentient, and affective relationship. The case of the anti-HEPP struggles in Turkey demonstrates that the rights and interests of nonhuman nature and "humans in nature" are not at odds. Instead, caring for one's own life and caring for the environment can be one and the same, so long as communities perceive themselves as part and parcel of the nonhuman world that surrounds them.

notes

1. Peter Münch, "Ein Tag am Meer" [One day at sea], *Süddeutsche Zeitung*, September 11, 2020, https://www.sueddeutsche.de/panorama/israel-westjordanland-1.5027731 (translated from German by the author).
2. For an exception, see Ingrid Leman Stefanovic, ed., *The Wonder of Water: Lived Experience, Policy, and Practice* (Toronto: University of Toronto Press, 2020).
3. Thiago B. A. Couto and Julian D. Olden, "Global Proliferation of Small Hydropower Plants—Science and Policy," *Frontiers in Ecology and the Environment* 16, no. 2 (2018): 91–100.
4. Erdem Evren, *Bulldozer Capitalism: Accumulation, Ruination, and Dispossession in Northeastern Turkey* (Oxford: Berghahn Books, 2022); Fiachra Gibbons and Lucas Moore, "Turkey's Great Leap Forward Risks Cultural and Environmental Bankruptcy," *The Guardian*, April 29, 2011, http://www.theguardian.com/world/2011/may/29/turkey-nuclear-hydro-power- development.
5. See Turkish Water Assembly, "HEPP's, Dams and the Status of Nature in Turkey," 2011, https://www.dogadernegi.org/wp-content/uploads/2015/10/HEPP_DAM_Turkey.pdf. And see Oğuz Kurdoğlu, "Expert-Based Evaluation of the Impacts of Hydropower Plant Construction on Natural Systems in Turkey," *Energy & Environment* 27, nos. 6–7 (2016): 690–703.
6. Çağıl Kasapoğlu, "100 Kadın: Karadeniz'in Dereleri İçin Direnen Kadınları" [100

women: The Black Sea women resisting for their rivers], *BBC Turkish*, October 25, 2013, http://www.bbc.com/turkce/haberler/2013/10/131025_findikli_kadinlar_ kasapoglu. See also Radikal, "Havva Ana Yalnız Değil: Çevreyi Paçavra Ettunuz... Biz de Sizi Edeceğuz" [Mother Havva (Eve) Is Not Alone: You Have Messed with the Environment... We Will Mess with You]," Kuzey Ormanları Savunması, July 19, 2015, https://kuzeyormanlari.org/2015/07/20/havva-ana-yalniz-degil-cevreyu-pacavra-ettunuz-biz-da-sizi-edeceguz/; Özge Yaka, "A Feminist-Phenomenology of Women's Activism against Hydropower Plants in Turkey's Eastern Black Sea Region," *Gender, Place and Culture* 24, no. 6 (2017): 868–89.

7. Between 2008 and 2018, there were 203 HEPPs built in the region, and 123 more were in the project phase.

8. Celene Krauss, "Challenging Power: Toxic Waste Protests and the Politicization of White, Working-Class Women," in *Community Activism and Feminist Politics: Organizing across Race, Class and Gender*, ed. Nancy A. Naples (New York: Routledge, 1998), 129–50. See also D. M. Prindeville, *On the Streets and in the State House: American Indian and Hispanic Women and Environmental Policymaking in New Mexico* (New York: Routledge, 2004). And see Shannon Elizabeth Bell, *Our Roots Run Deep as Ironweed: Appalachian Women and the Fight for Environmental Justice* (Urbana: University of Illinois Press, 2013).

9. The gender difference in the ways men and women frame their opposition to HEPPs and enact their political agencies resonate with the feminist political ecology literature, which stresses "real, not imagined, gender differences in experiences of, responsibilities for, and interests in 'nature' and environments" that are "not rooted in biology per se." D. Rocheleau, B. Thomas-Slayter, and E. Wangari, *Feminist Political Ecology: Global Issues and Local Experiences* (London: Routledge, 1996), 2. They are instead rooted in the gendered organization of everyday life and material practices, especially in the gendered division of labor. The gendered division of labor in the East Black Sea region, for instance (i.e., agricultural work is perceived and performed as an extension of housework, and it is women who work in the fields planting, fertilizing, pruning, and harvesting tea plants), provides the material conditions for women to interact with river waters in a routine and habitual manner. As the fields (as well as the houses) are located by rivers within the long, deep valleys that run through the Kaçkar (Pontic) Mountains to the Black Sea, agricultural work puts women in constant connection with rivers. For a detailed discussion, see Özge Yaka, *Fighting for the River: Gender, Body, and Agency in Environmental Struggles* (Oakland: University of California Press, 2023).

10. Özge Yaka, "Gender and Framing: Gender as a Main Determinant of Frame Variation in Turkey's Anti-Hydropower Movement," *Women's Studies International Forum* 74 (2019): 154–61.

11. For more examples and a detailed discussion, see Yaka, *Fighting for the River*.

12. Diana Coole, "Rethinking Agency: A Phenomenological Approach to Embodiment and Agentic Capacities," *Political Studies* 53, no. 1 (2005): 135.

13. Lifeworld is a phenomenological concept that originated in Husserl's work but is employed by many others, including Jürgen Habermas. It designates the given, self-evident world of lived experience.

14. Donna Haraway, *Simians, Cyborgs, and Women: The Reinvention of Nature* (New York: Routledge, 1991). See also Sarah Whatmore, *Hybrid Geographies: Natures Cultures Spaces* (London: Sage Publications, 2002); Bruno Latour, *Reassembling the Social: An Introduction to Actor-Network Theory* (Oxford: Oxford University Press, 2007).

15. Luke Whitmore, *Mountain, Water, Rock, God* (Oakland: University of California Press, 2019).

16. Arturo Escobar, "Sustainability: Design for the Pluriverse," *Development* 54, no. 2

(2011): 137–40. See also Soren C. Larsen and Jay T. Johnson, *Being Together in Place Indigenous Coexistence in a More Than Human World* (Minneapolis: University of Minnesota Press, 2017).

17. P. Mohai, D. Pellow, and J. T. Roberts, "Environmental Justice," *Review of Environment and Resources* 34 (2009): 405–30.

18. Nicholas P. Low and Brendan Gleeson, *Justice, Society and Nature: An Exploration of Political Ecology* (London: Routledge, 1998). See also Brain Baxter, *A Theory of Ecological Justice* (New York: Routledge, 2005).

19. Özge Yaka, "Rethinking Justice: Struggles for Environmental Commons and the Notion of Socio-Ecological Justice," *Antipode* 51, no. 1 (2019): 353–72. See also Özge Yaka, "Justice as Relationality: Socio-Ecological Justice in the Context of Anti-Hydropower Movements in Turkey," *DIE ERDE–Journal of the Geographical Society of Berlin* 151, nos. 2–3 (2020): 167–80.

20. Muhammet Kaçar, "Arılı Vadisinde Yola Oturan Kadınlar Bilirkişi Heyetinin Yolunu Kesti" [The women block the road to in the Arılı Valley]," *Hürriyet*, July 18, 2017, https://www.hurriyet.com.tr/arili-vadisinde-yola-oturan-kadinlar-bilirkisi-40523696.

Isle of One

J. Drew Lanham

Sitting here insular.

Really though, I'd like to exercise the superlative of that state, to be on more of an island. Maybe a smaller island. A more away island; somehow be further removed from all the hate and hypocrisy flooding every island. Sitting here looking at the ocean water churn muddy brown in St. Helena Sound, sending enough sand shoreward each second for me to push my head shoulder deep into it until the successive comings and goings of the tide bury me like all the flotsam and shells that have ever come before.

Pour upon and bury me until the deafening report of gunfire killing black people praying and dog-whistled sound bitten lies of history and not hate cease; until the empty prayers for school children shot cowering under desks and heedless hopes of do-nothings who can't even agree that the sun rises and sets, are muffled into surf and gull laughter and dolphin blowing.

I can only escape the din for a moment while looking at some bird or wandering away on my own to lose now in what's better to maybe never come. In the second of seeing feather flashing or searching for the answer on a lettered olive shell, I don't have to look at anything else.

I don't have to think, on that island

But then I come back to the mainland of reality. I blink back from staring at wing beats or beak shape or plumage hue. My fool's exercise in false escapism gives me just enough clarity and sanity to exist until it's my turn or someone I know–because I can still hear it from this island. It is close to every one of us everywhere. Whether here or the center of mainland connected. No matter where I am,

we are—It is here; the unceasing headwind—the expanse of featureless sea with no horizon to hope for except the one we hope for.

Meanwhile,

I drown for the ancestors brave enough to throw themselves into the abyss rather than endure all this.

I swim strong for the ancestors brave enough to endure the slavehold, breathe fetid air and endure all this; the can't see to can't see, the whip. the rape, the Jim Crow, the redlines, the gerrymanders, the side-eyes, the crosses burned in the night, the lynch ropes stretched, the supreme court cases ruled for and against, the spittle for god-fearing people aimed at me, the back door way to my colored-only seat, the open caskets that should have stayed closed, the water too dirty to drink, the water in the public pool where I was not allowed to dip, the ocean segregated so I could not see from whence I just came a few years hence, the doors locking when I walk by, the purses clutched in elevators, the credit scores not good enough, the pseudo woke piety of so-called progressives who don't want me in their back yards, the MLK Day content of their character paper-thin excuses to not see my color so you can continue your prejudice.

Meanwhile,

I can hear it from every place I have ever been or could
think of being. Wildness is no cure for inhumanity's blight of
cruelty exacted because of skin color or who we love or the
language we speak. I don't think there's a place to flee. There's
no away that's far enough away any more to ever be insulated
or safe.

Maybe hopelessness is an island unto itself. Maybe hope
is the sandbar somewhere out there where whimbrel and
oystercatchers rest. I need to find that place. That peace of
isolation. That peace of reality estate.

Permissions

Acknowledgments

Our gratitude runs deep for the community of kin who made this series possible. Strachan Donnelley, the founder of the Center for Humans and Nature, was animated and inspired by big questions. He liked to ask them, he enjoyed following the intellectual and actual trails where they might lead, and he knew that was best done in the company of others. Because of this, and because Strachan never tired of discussing the ancient Greek philosopher Heraclitus, who was partial to Fire, we think he would be pleased by the collective journey represented in *Elementals*. One of Strachan's favorite terms was "nature alive," an expression he borrowed from the philosopher Alfred North Whitehead. The words suggest activity, vivaciousness, generous abundance—a world alive with elemental energy: Earth, Air, Water, Fire. We are a part of that energy, are here on this planet because of it, and the offering of words given by our creative, empathic, and insightful contributors is one way that we collectively seek to honor *nature alive*.

A well-crafted, artfully designed book can contribute to the vitality of life. For the mind-bending beauty of the cover design, cheers to Mere Montgomery of LimeRed; she is a delight to work with and LimeRed an incredible partner in bringing to visual life the Center for Humans and Nature's values. For an eye of which an eagle would be envious, a thousand blessings to the deft manuscript editor Katherine Faydash. For the overall style and subtle touches to be experienced in the page layout and design, we profoundly thank Riley Brady. We also wish to thank Ronald Mocerino at the Graphic Arts Studio Inc. for his good-natured spirit and attention to our printing needs, and Chelsea Green Publishing for being excellent collaborators in distribution and promotion.

Thank you to our colleagues at the Center for Humans and Nature, who are elemental forces in their own rights, including our president Brooke Parry Hecht, as well as Lorna Bates, Anja Claus, Katherine Kassouf Cummings, Curt Meine, Abena Motaboli, Kim Lero, Sandi Quinn, and Erin Williams. Finally, this work could not move forward without the visionary care and support of the Center for Humans and Nature board, a group that carries on Strachan's legacy in seeking to understand more deeply our relationships with *nature alive*: Gerald Adelmann, Julia Antonatos, Jake Berlin, Ceara Donnelley, Tagen Donnelley, Kim Elliman, Charles Lane, Thomas Lovejoy, Ed Miller, George Ranney, Bryan Rowley, Lois Vitt Sale, Brooke Williams, and Orrin Williams.

—Gavin Van Horn and Bruce Jennings
series coeditors

I am particularly grateful to Bruce Jennings and Gavin Van Horn for having invited me to participate in this exciting project. It was an honor to have this opportunity of working with two such exceptional writers and editors. I also thank each contributor, upon whom the richness of the water volume depends. Finally, I appreciate, as always, my husband Michael, who roots me, supports me, and fulfills me, as well as my family, who keeps me whole.

—Ingrid Leman Stefanovic, editor

Contributors • volume iii

Clifford Atleo (he/him) is Assistant Professor at the School of Resource and Environmental Management at Simon Fraser University, where he teaches and researches Indigenous governance, political economy, and resource management. He is Nuu-chah-nulth and Tsimshian.

Elizabeth Bradfield (she/her) is the author of five collections of poetry, most recently *Toward Antarctica*, which uses *haibun* and photographs to query her work as a naturalist in Antarctica, and *Theorem*, a collaboration with artist Antonia Contro. She is also coeditor of *Cascadia Field Guide: Art, Ecology, Poetry* and

Photo by Lisa Sette

Broadsided Press: Fifteen Years of Poetic/Artistic Collaboration, 2005–2020. Bradfield's work has appeared in the *New Yorker, The Atlantic, Poetry,* and *The Sun*, and her honors include the Audre Lorde Prize and a Stegner Fellowship. Based on Cape Cod, Liz works as a naturalist, teaches at Brandeis University, and runs Broadsided Press. www.ebradfield.com

Nickole Brown (she/her) is the author of *Sister* and *Fanny Says*. She lives in Asheville, North Carolina, where she volunteers at several animal sanctuaries. *To Those Who Were Our First Gods*, a chapbook of poems about these animals, won the 2018 Rattle Prize, and her essay-in-poems, *The Donkey Elegies,* was published by Sibling Rivalry Press in 2020. She's the President of the Hellbender Gathering of Poets, an annual environmental literary festival in Black Mountain, North Carolina.

Hannah Close is a writer, photographer, curator, and researcher. She is currently making a documentary called *Islandness in the Anthropocene* and completing an MA in engaged ecology at Schumacher College with a dissertation project exploring archipelagic poetics and relational metaphors.

Geffrey Davis (he/him) is the author of three books of poems, most recently *One Wild Word Away* (BOA Editions, 2024). His second collection, *Night Angler,* won the James Laughlin Award from the Academy of American Poets, and his debut, *Revising the Storm,* won the A. Poulin Jr. Poetry Prize. Davis lives in the Ozarks and teaches for the Program in Creative Writing & Translation at the University of Arkansas. Raised by the Pacific Northwest, he is also a core faculty member of the Rainier Writing Workshop and serves as Poetry Editor for *Iron Horse Literary Review.*

Margo Farnsworth (she/her) writes and speaks about biomimicry and nature-based solutions to professionals interested in innovative business practices. When she's not writing, she helps organizations connect with stakeholders and enact their strategic plans. Her book, *Biomimicry & Business: How Companies Are Using Nature's Strategies to Succeed,* stories how five companies discovered biomimicry and are partnering with nature. Her work also appears in *Wildness: Relations of People & Place* as well as various magazines and blogs. She is working on a biomimicry book series for children to introduce them to the ways they can learn from our wild neighbors.

Forrest Gander, born in the Mojave Desert, lives in California. A translator and writer with degrees in geology and literature, he's received the Pulitzer Prize, Best Translated Book Award, and fellowships from the Library of Congress, Guggenheim, and as a United States Artists Rockefeller Fellow. His book *Twice Alive* focuses on human and ecological intimacies. New Directions published his long poem on the desert, *Mojave Ghost,* in 2024.

Photo by Denise Toombs, courtesy of the YWCA

Joy Harjo (she/her), the 23rd US Poet Laureate and member of the Muscogee Nation, is the author of ten books of poetry, several plays, children's books, two memoirs, and seven music albums. Her honors include Yale's 2023 Bollingen Prize for American Poetry, the National Book Critics Circle Ivan Sandrof Lifetime Achievement Award, the Ruth Lilly Prize from the Poetry Foundation, a Guggenheim Fellowship, and a Tulsa Artist Fellowship. She is Chancellor of the Academy of American Poets and Chair of the Native Arts & Cultures Foundation, and is the inaugural Artist-in-Residence for the Bob Dylan Center in Tulsa, Oklahoma, where she lives.

Lyanda Fern Lynn Haupt (she/her) is a writer and naturalist based in the Pacific Northwest. Her work explores the beautiful, complicated connections between humans and the wild, natural world. Lyanda is the award-winning author of several books, including *Crow Planet, Mozart's Starling,* and *Rooted: Life at the Crossroads of Science, Nature, and Spirit.*

Bruce Jennings (he/him) teaches and writes on ethical and social issues in health care at Vanderbilt University. He is Developmental Editor for Humans & Nature Press Books and Senior Fellow at the Center for Humans and Nature. He is author of several books and many articles in the fields of biomedical ethics, public health, and ecological ethics. Among his books is *Ecological Governance: Toward a New Social Contract with the Earth* (2016).

J. Drew Lanham (he/him) is a poet, writer, and ornithology professor at Clemson University. He is Poet Laureate of Edgefield, South Carolina, and the author of *Sparrow Envy: Field Guide to Birds and Lesser Beasts, Joy Is the Justice We Give Ourselves,* and *The Home Place: Memoirs of a Colored Man's Love Affair with Nature.* Drew's work centers on ethnic perspectives of wildness and nature conservation as activism. He is a 2022 MacArthur Fellow and an avid bird watcher who lives on a 45-acre farm in the Dark Corner of South Carolina.

Raised in the Ecuadorian Amazon and Andes among volcanoes, **Lisa María Madera's** (she/her) writing explores the intersections of religion, nature, and extractivist economies through the lens of (neo)colonial myth and history. As an educator and social entrepreneur, Lisa works to foster compassionate and resilient communities by providing extraordinary encounters with nature and her many creatures. She is currently writing a memoir entitled *The Covid Chronicles: Lessons from Pacha Mama in the Face of Despair.*

Marzieh Miri (she/her) is a Toronto-based multidisciplinary artist who was born and raised in Iran. She has an MFA in documentary media from Toronto Metropolitan University and a master's in architecture. Her research and creative practice explore the notions of place, land, and environment through photographic mediums. She is especially interested in practice-based and sensorial approaches investigating humans and their environment as a united existence. Winner of Ontario Art Council grants, she has exhibited her art projects in Iran, France, Austria, and Canada. She has also worked as an architect, professor, writer, and critic and has published in international journals and presented at conferences and courses in Canada, England, and Iran.

Kathleen Dean Moore is an environmental philosopher and the author of award-winning books about our moral and emotional connections to the wild, reeling world. Formerly Distinguished Professor at Oregon State University, she left academia to write and speak about the moral urgency of climate action. Her climate ethics books are *Moral Ground, Great Tide Rising, Earth's Wild Music,*

Bearing Witness, and *Take Heart: Encouragement for Earth's Weary Lovers.* Kathleen writes from the Willamette Valley in Oregon and from a cabin built over the tide in wilderness Alaska.

Martin Lee Mueller (he/him), PhD, is a philosopher, storyteller, cofounder of the Norwegian Deep Ecology Festival Motstrøms, and author of the award-winning *Being Salmon, Being Human.* His recent research involves action-oriented, community-based approaches to transformative education in the face of ecological grief. He is an affiliated scholar with the University of Oslo and the University of South-East Norway in Bø, Telemark. The water bodies he is most drawn to are the Oslo Fjord and the Akerselva River, which once suffered a fatal chlorine spill but now sees salmon spawn again every November, right in the heart of his city.

Mark Riegner (he/him) has a PhD in ecology and evolution from SUNY Stony Brook and taught in the Environmental Studies Program at Prescott College in Arizona from 1988 until his recent retirement in 2023. His courses offered a foundation in ecology and animal biology and combined theory with an experiential emphasis, particularly in his field courses in Mexico, Costa Rica, and East Africa. His research interests embrace phenomenological approaches to studying patterns in nature, and he has published on reading the subtle qualities of landscapes, on the relationship of form and pattern in mammals, and on morphology and color-pattern evolution in birds.

Photo by
Marguerite Legros

Anna Selby (she/her) is a poet and naturalist. Her most recent chapbook, *Field Notes,* was one of the LRB Bookshop's Bestsellers for two years running and was an *Irish Times* Book of the Year. She is Lecturer at Schumacher College, teaching engaged ecology; is working on a PhD on empathy, ecology, and plein air poetry at Manchester Metropolitan University, and was editor and cofounder of environmental, feminist publisher Hazel Press and one of the judges for the 2022 Ginkgo Prize for Ecopoetry.

Ingrid Leman Stefanovic (she/her) is an author, philosopher, and consultant in environmental and institutional change management. Formerly Dean of the Faculty of Environment at Simon Fraser University, Vancouver, Canada, she is Professor Emerita, Department of Philosophy, University of Toronto, where her research and teaching focused on how values affect public policy, planning, and environmental decision making. Key books include *Safeguarding Our Common Future* as well as edited volumes *The Natural City, The Wonder of Water: Lived Experience, Policy and Practice,* and *Ethical Water Stewardship.*

CD Wright was a major innovative poet who was born in Arkansas. Her many distinctive books, widely translated, influenced the course of documentary poetics and twenty-first century political poetry while being idiosyncratic and marked by her Southern roots. In her obituary, the *New York Times* asserted that she "constituted a school of precisely one." Her legacy continues with new translations, the annual 40K CD Wright Award from the Foundation for Contemporary Arts, the annual CD Wright Memorial Lecture at Brown University, the annual CD Wright Southern Women's Poetry Conference, and with forthcoming critical books on her work and a biography.

Robert Wrigley (he/him) has published twelve books of poems, most recently *The True Account of Myself as a Bird*. He lives in Idaho and spends as much time as he can in Italy.

Özge Yaka (she/her) received her PhD in sociology from Lancaster University. She is currently based at Freie Universität Berlin, working at the intersection of human geography, environmental humanities, and gender studies.